Feedback and Dynamic Control
of Plasmas

AIP Conference Proceedings

Series Editor: Hugh C. Wolfe

Number 1

Feedback and Dynamic Control of Plasmas

Editors

T. K. Chu

Plasma Physics Laboratory, Princeton University

and

H. W. Hendel

Plasma Physics Laboratory, Princeton University
(on leave from RCA Laboratories, Princeton, N.J.)

American Institute of Physics
New York   1970

Library of Congress Catalog Card
Number 70-141596

American Institute of Physics
335 East 45 Street
New York, N.Y. 10017

Printed in the United States of America

PREFACE

This volume comprises the manuscripts presented at the Symposium on Feedback and Dynamic Control of Plasmas, held in Princeton in June, 1970.

Adequate confinement in controlled thermonuclear fusion reactors may depend critically on our ability to control the losses caused by instabilities. Methods of feedback and dynamic control have led to significant experimental results only during the last few years, although some theoretical groundwork had been laid earlier. The present volume summarizes the current state of feedback and dynamic-stabilization work in plasmas.

Feedback experiments have been performed in many devices, especially linear ones. Results have been in good agreement with theory. However, the reported work is restricted to the region of marginal instability and coherent oscillations, and generally to the presence of only one mode. Dynamic stabilization is predicted to stabilize many modes simultaneously, but the reported experimental results are limited and their interpretation is difficult. Thus, with promising initial results, much room is left for future contributions.

It is hoped that these Proceedings will contribute to the achievement of controlled thermonuclear fusion by emphasizing nonstatic control methods, which have been so effective in other fields of technology.

Drs. D. Pfirsch, M. N. Rosenbluth, and J. B. Taylor, of the Symposium Advisory Committee, and Drs. H. P. Furth and F. L. Ribe were instrumental in the arrangement of the Symposium program. We thank many staff members of the Plasma Physics Laboratory for their help in the preparation of these Proceedings. Dr. M. B. Gottlieb has made possible both the Symposium and these Proceedings.

T. K. C.
H. W. H.

v

CONTENTS

Part I

FEEDBACK CONTROL

GENERAL FEEDBACK THEORIES AND FEEDBACK THEORIES FOR ELECTROSTATIC MODES

1.1 FEEDBACK STABILIZATION OF A VLASOV PLASMA

M. Cotsaftis

Association Euratom-CEA, Centre d'Études Nucléaires
Fontenay-aux-Roses (France)

ABSTRACT

The complete Maxwell-Vlasov equations together with feed-back are analyzed in the quasi-homogeneous case, using the Nyquist criterion. A class of feedback loops, satisfying sufficient conditions for stability but depending on the wave number k, is found. If the instability satisfies certain properties, a subclass of feedback loops, independent of the wave number, can be constructed. The classi-cal amplification-delay feedback, generally used in experiments, does not belong to the subclass; and therefore designing an adapted feedback loop is of importance. Applications of the general theory to the low-frequency electrostatic microinstability and the double beam instability are given.

I. INTRODUCTION

Many recent plasma experiments have shown an interesting stabilizing effect when feedback is added to the device (Arsenin, 1970; Keen and Fletcher, 1970; Simonen, Chu, and Hendel, 1969). This effect could be expected to originate from the structure of the complete Maxwell-Vlasov equations. It is our purpose to show that for this feedback-loop system, sufficient conditions for stability, easily satisfied by a large class of feedbacks, can be found.

II. THEORY

Let the loop diagram represent the Maxwell-Vlasov equations with a feedback system (Fig. 1). We restrict ourselves to the quasi-homogeneous electrostatic case first. Here $K = K(p, k)$ is the transfer function of the feedback, including detection and transformation of the signals ($K = 1$, without feedback). A pertur-bation then obeys the linearized equations:

$$\frac{D}{Dt}\, \delta f = - \delta \vec{E}^* \cdot \vec{\nabla}_v f_o \; , \tag{1}$$

$$\mathrm{div}\, \delta \vec{E} = \omega_p^2 \int \delta f d^3 \vec{v} \; , \tag{2}$$

$$\delta \vec{E}^* = K \delta \vec{E} \; , \tag{3}$$

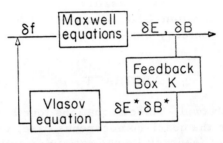

Fig. 1

where $\frac{D}{Dt}$ is the total derivative along the unperturbed orbit. The dispersion relation is:

$$D(p,k) \equiv k^2 + K\,(p,k)\; G\,(p,k) = 0 \tag{4}$$

where $p = \gamma + i\omega$ is the Laplace variable. A useful property of G is:

$$\lim_{|p| \to \infty} G\,(p,k) = 0 \tag{5}$$

Hence,

$$\lim_{|p| \to \infty} D\,(p,k) = k^2 > 0. \tag{6}$$

Applying the Nyquist criterion[*] gives two sufficient conditions for stability:

$$\lim_{|p| \to \infty} D\,(p,k) > 0 \tag{6a}$$

$$\mathrm{Re}\left\{ pK\,(p,k)\; G\,(p,k) \right\} > 0; \;\; \gamma \gtrless 0, \; -\infty < \omega < +\infty. \tag{6b}$$

These conditions are obtained from an analysis in the (ReD, ImD) plane of the half positive semi circle with radius going to infinity and closed by a parallel at a distance γ from the ω axis.

---

[*]Handbook of Automation, Computation and Control, Vol. II, J. Wiley and Sons, p. 21-09 (1959). Eds. E.M. Grabbe, S. Ramo and D.E. Wooldridge

Without feedback, (6a) is satisfied; in an unstable case (6b) is not satisfied everywhere. Writing $G = X + iY$, we obtain:

$$X_o \equiv \gamma X - \omega Y < 0 \tag{7}$$

in at least one interval of the ω-axis. With feedback $K = U + iV$, stability will be restored if U and V can be found such that both (6a) and (6b) are satisfied. If we consider only that class of K for which $\lim_{|p| \to \infty} K \to 0$ (the class of high-frequency oscillations), (6a) is trivially satisfied; and (6b) requires

$$UX_o - VY_o > 0; \quad \gamma \gtreqless 0, \quad -\infty < \omega < +\infty \tag{8}$$

where $Y_0 \equiv \omega X + \gamma Y$. This problem has an infinite number of possible solutions. A specific choice depends on knowledge of the "open loop" and the requirements on the output properties, as usual in system design.

III. APPLICATION

If first Landau or dissipative effects can be neglected, as for flute or trapped-particle instabilities, $G(p, k)$ is a real function of its arguments. For small γ,

$$G(i\omega + \gamma, k) = X(\omega, k) - i\gamma \frac{\partial X}{\partial \omega} . \tag{9}$$

Assuming a similar property for K ($U = U(\omega, k)$, $V = -\gamma \frac{\partial U}{\partial \omega}$, we obtain from (8)

$$\frac{\partial}{\partial \omega}(\omega UX) = U \frac{\partial}{\partial \omega}(\omega X) + \omega X \frac{\partial U}{\partial \omega} > 0; \tag{10}$$

while Eq. (7) requires

$$\frac{\partial}{\partial \omega}(\omega X) < 0 \tag{11}$$

in some ω-interval (for the unstable case). The possible solutions of (10) and (11) are:

$$U = \frac{\int \phi^2(\omega') \, d\omega'}{\omega X(\omega, k)}, \tag{12}$$

where ϕ^2 is any convenient positive function. However, it is evident that this solution is k-dependent, i.e., adapted for stabilizing one mode only. For a specific plasma, one thus would need

as many loops as there are unstable modes. This result is inferior
to that obtained from dynamical stabilization (Cotsaftis, 1968) by
which <u>all</u> modes can be stabilized. Nevertheless, assuming
further properties, we obtain, for instance:

$$\omega X \ge - M_o \frac{\partial}{\partial \omega} (\omega X) \ge - M_1 \tag{13}$$

in the considered ω-domain, M_o and M_1 being two positive functions
of k, such that $m \ge M_1 / M_o$ (m a positive number). A solution of
U is then:

$$U' + mU + \phi^2 (\omega) = 0, \tag{14}$$

and is independent of the wave number k. In that case, even
knowing the poles of the open-loop system in unnecessary. For a
double-stream instability, $m = 3\sqrt{3}/2$. From (14) we can obtain
U for a fixed ϕ.

  More generally, when Landau effects are important, as for the
universal instability, (Rosenbluth, 1965) Eqs. (8) and (7) require
respectively,

$$- (\omega X V + \omega U Y) + \gamma \left\{ \frac{\partial}{\partial \omega} (\omega U X) - \frac{\partial}{\partial \omega} (\omega V Y) \right\} > 0 \tag{15}$$

and

$$-\omega Y + \gamma \frac{\partial}{\partial \omega} (\omega X) < 0 . \tag{16}$$

In the limit $\gamma \to 0$, a possible solution to (15) is:

$$V (\omega) = 0, \quad U = U(\omega, k), \quad \text{sign } U = -\text{sign} (\omega, y) \tag{17}$$

with K being still a real function of its arguments. For a double-
stream instability, the corresponding U is given on Fig. 2. Apart

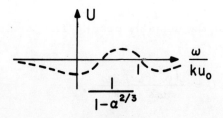

Fig. 2

from the sign conditions (17), the shape of U can be freely modeled
to lower the oscillation level and to damp out the noise,   for
instance. A more general low-frequency dispersion relation
including flute, trapped-particle, and universal instabilities can be
formulated as well. However, this solution is still k-dependent.
To get an approximate result, the universal instability alone has
been studied and can be stabilized for any k in the interval (k_{min},
k_{max}) by the K shown in Fig. 3. It is worthwhile remarking that
the usual amplification-delay feedback largely used in experiments

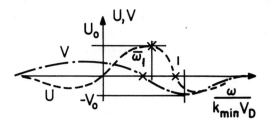

Fig. 3

with $K = 1 + K_0 \exp(-p\tau)$, K_0, $\tau > 0$ may be k-dependent and does
not appear to be a sufficiently general one, because (7) is only
satisfied when

$$K_0 > 1, \quad \omega\tau \simeq (2n + 1)\pi \tag{18}$$

which severely restricts ω. Therefore, it is expected to be useful for
only one instability mode, in contrast to the class of feedbacks
discussed here. These results can be extended to the electromagnetic case as well with the full Maxwell equations.

IV. SUMMARY

For a Vlasov plasma there exists an infinite class of feedbacks
giving "mode stability", and a more interesting infinite subclass of
them producing global stability. These feedbacks can furthermore be
be modeled according to other criteria. The usual amplification-
delay feedback appears rather restrictive, and choosing a feedback
of the second class would be preferable. Finally, the problem of
operating the feedback, supposing one can design it, is left open.
A possible way for sending back into the plasma the transformed
signal could be the non-linear conversion of waves in inhomogeneous
media (Wong, Baker, and Booth, 1970). Therefore extending these
results to the inhomogeneous case would be of interest.

1.2 PLASMA CONFINEMENT BY LOCALIZED
FEEDBACK CONTROLS

P. K. C. Wang

School of Engineering and Applied Science
University of California, Los Angeles

ABSTRACT

The problem of confining a plasma in a specified bounded spatial domain by feedback-control forces applied over an outer shell of the domain is considered. Using a two-component fluid model for the plasma, various forms of feedback controls are derived by minimizing an instantaneous weighted mean outflow rate over a subregion of the shell. Effects of some of the derived feedback controls on plasma confinement are determined by experiments performed on a two-dimensional computer-simulated plasma.

I. INTRODUCTION

The problem of confining a plasma in a specified bounded spatial domain by means of external time-varying electromagnetic fields has been explored using the Vlasov model with certain simplifying assumptions (Wang, 1969; Wang and Janos, 1970). The results are in the form of feedback controls depending on the instantaneous outflow averaged over the boundary surface of the confinement region. In physical situations, it is difficult to generate time-varying electromagnetic fields with specified spatial variations for controlling the plasma motions. Localizations of the controls to certain portions of the plasma are usually necessary for their physical realization. The objective here is to derive and compare various forms of feedback controls for reducing the plasma leakage from a specified spatial domain when they are localized to an outer shell of the domain.

II. MATHEMATICAL MODEL

Let D, the plasma confinement region, be a specified closed bounded domain with boundary surface ∂D in the three-dimensional space. Let D_s represent an outer shell of D with nonzero thickness formed by the union of subregions (cells) D_j ($D_s = \bigcup_{j=1}^{k} D_j$). Assume that the plasma is initially confined in D at time $t = 0$. The problem is to control its leakage or outflow from D by applying suitable control (manipulatable) forces localized to D_s. The localization of control forces to an outer shell of D is motivated from the fact that any particle must penetrate D_s before escaping from D. Also, for controls in the form of electromagnetic fields, it is difficult to attain deep penetration of the fields into the plasma to achieve effective control.

The determination of feedback controls is based on the observation that if, at any time, the plasma outflow from D over some future time duration can be extrapolated by means of a suitable mathematical model, then the control forces can be chosen in such a way to reduce the extrapolated (future) outflow. The resulting control is, in effect, an <u>anticipative</u> action. To illustrate the basic ideas, a two-component fluid model for the plasma described by the following equations will be used to obtain specific results:

$$m_e n_e d\vec{v}_e/dt = -\vec{\nabla}p_e - en_e (\vec{E} + \vec{v}_e \times \vec{B}) + \vec{S}_e \quad , \tag{1}$$

$$m_i n_i d\vec{v}_i/dt = -\vec{\nabla}p_i + Zen_i (\vec{E} + \vec{v}_i \times \vec{B}) + \vec{S}_i \quad , \tag{2}$$

where the subscripts "e" and "i" denote the electron and ion fluids respectively; n_e and n_i are number densities satisfying the continuity equations, and $\vec{S}_e, \vec{S}_i$ are momentum sources. Let $\vec{M}_e^j(t)$, $\vec{M}_i^j(t)$ and $\vec{M}_j(t)$ denote respectively the average momenta of the electron and ion fluids and a weighted average total momentum in cell D_j at time t defined by

$$\vec{M}_e^j(t) = \int_{D_j} m_e n_e(t,\vec{r})\vec{v}_e(t,\vec{r})dD_j, \quad \vec{M}_i^j(t) = \int_{D_j} m_i n_i(t,\vec{r})\vec{v}_i(t,\vec{r})dD_j \quad , \tag{3}$$

$$\vec{M}_j(t) = \alpha_e \vec{M}_e^j(t) + \alpha_i \vec{M}_i^j(t), \quad \alpha_e, \alpha_i \geq 0 \quad , \quad \alpha_e + \alpha_i = 1 \quad . \tag{4}$$

Let $\vec{\eta}_j$ denote an average outward unit normal associated with D_j (e.g., $\vec{\eta}_j$ may correspond to the outward unit normal averaged over the outer boundary surface of D_j). Straightforward computations show that $\vec{\eta}_j \cdot \vec{M}_j(t)$, the outward component of average momentum of the plasma in any cell D_j, satisfies:

$$\frac{d}{dt}(\vec{\eta}_j \cdot \vec{M}_j) = -\vec{\eta}_j \cdot \int_{D_j} \vec{\nabla}(\alpha_e p_e + \alpha_i p_i)dD_j + e\vec{\eta}_j \cdot \int_{D_j} (\alpha_i Zn_i - \alpha_e n_e)\vec{E}\, dD_j$$

$$+ e\vec{\eta}_j \cdot \int_{D_j} (\alpha_i Zn_i \vec{v}_i - \alpha_e n_e \vec{v}_e) \times \vec{B}\, dD_j + \vec{\eta}_j \cdot \int_{D_j} (\alpha_e \vec{S}_e + \alpha_i \vec{S}_i)\, dD_j \quad ,$$

$$j = 1, \ldots, k \quad . \tag{5}$$

III. DERIVATION OF FEEDBACK CONTROLS

From Eq. (5), various forms of feedback controls for each cell can be derived by minimizing the instantaneous weighted mean outflow rate associated with each cell. Note that minimizing the outflow rate leads to, in effect, an anticipative control.

A. Electric Field Control: Let $\vec{E} = \vec{E}_p + \vec{E}_c$, where $\vec{E}_p$ and $\vec{E}_c$ are the electric fields due to the plasma and the manipulatable sources (control) respectively. Let $\vec{E}_c$ be uniform within each D_j. (Denote $\vec{E}_c$ in D_j by $\vec{E}_c^j$). The second integral in Eq. (5) can be rewritten as:

$$e\vec{\eta}_j \cdot \int_{D_j} (\alpha_i Z n_i - \alpha_e n_e)\vec{E}_p \, dD_j + \left\{ e \int_{D_j} (\alpha_i Z n_i - \alpha_e n_e) \, dD_j \right\} (\vec{\eta}_j \cdot \vec{E}_c^j). \quad (6)$$

From physical limitations, magnitude constraints of the form $|\vec{E}_c^j(t)| \leq E_{max}^j = $ constant for all t; j = 1, ..., k are imposed. In the presence of these constraints, the second term in Eq. (6) is minimized if $\vec{E}_c^j(t)$ takes on the form:

$$\vec{E}_c^j(t) = \begin{cases} -E_{max}^j \, (\text{Sign } h(t))\vec{\eta}_j & \text{if } h(t) \neq 0, \\ 0 & \text{if } h(t) = 0, \end{cases}$$

$$h(t) = \int_{D_j} (\alpha_i Z n_i(t, \vec{r}) - \alpha_e n_e(t, \vec{r})) \, dD_j, \quad j = 1, ..., k . \quad (7)$$

The above equation corresponds to a switching type of <u>nonlinear</u> feedback control whose sign depends on the weighted total charge in each cell D_j. Clearly, if $h(t) \neq 0$, such a control tends to reduce the weighted localized outflow $\vec{\eta}_j \cdot \vec{M}_j(t)$. Another possible form of feedback control is given by

$$\vec{E}_c^j(t) = \begin{cases} -E_{max}^j \, (\text{Sign } h(t))\vec{M}_j(t)/|\vec{M}_j(t)| & \text{if } h(t) \text{ and } |\vec{M}_j(t)| \neq 0, \\ 0 & \text{if } h(t) \text{ or } |\vec{M}_j(t)| = 0, \; j = 1, ..., k . \end{cases} \quad (8)$$

This control tends to drive $\vec{\eta}_j \cdot \vec{M}_j(t)$ toward zero instead of a negative direction as in the case of Eq. (7). In both cases, the switching discontinuities in $\vec{E}_c^j(t)$ may be replaced by smooth transitions. In particular, for small values of $|h(t)|$, one may approximate Eqs. (7) and (8) by (7'), $\vec{E}_c^j(t) = -G_j h(t)\vec{\eta}_j$, and (8'), $\vec{E}_c^j(t) = -G_j h(t)\vec{M}_j(t)$, respectively, where G_j are large positive constants.

B. <u>Momentum Source Control</u>: Let $\vec{S}_j(t)$ be a momentum source associated with D as defined by the last integral in Eq. (5). In the presence of constraints: $|\vec{S}_j(t)| \leq S^j_{max}$ for all t, j = 1, ..., k, the following feedback controls analogous to Eqs. (7) and (8) can be established:

$$\vec{S}_j(t) = -S^j_{max}\vec{\eta}_j \quad , \quad (7''); \text{and } \vec{S}_j(t) = \begin{cases} -S^j_{max}\vec{M}_j(t)/|\vec{M}_j(t)| & \text{if } |\vec{M}_j(t)| \neq 0, \\ 0 & \text{if } |\vec{M}_j(t)| = 0, \end{cases}$$

$$j = 1, ..., k \quad . \quad (8'')$$

<u>Remark</u>: With the foregoing approach, other forms of feedback controls such as pressure and magnetic field controls can be established. Note that this approach does not rely on any detailed information or assumptions pertaining to the physical mechanism of plasma leakage from D.

IV. COMPUTER EXPERIMENT

Due to the difficulties in obtaining sharp estimates of the improvement in plasma confinement for the derived feedback controls directly from a priori estimates of the

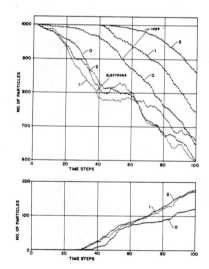

Fig. 1. Upper figure: Number of ions and electrons in D vs. time. Lower figure: Number of electrons absorbed at the walls vs. time. Curve 0: without control; Curve 1: with feedback control given by Eq. (7); Curve 2: with feedback control given by Eq. (7'). Time scale: 1 step = $0.05 \, \omega_{ce}^{-1}$; ω_{ce} —electron cyclotron frequency corresponding to $B_{zo} = 0.5$ webers/m^2; mass ratio: $m_i/m_e = 18.3$.

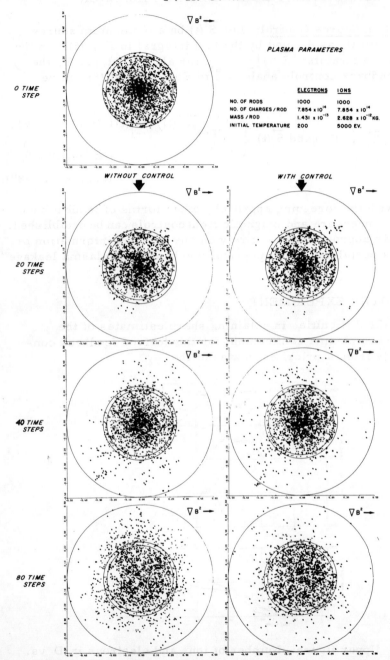

Fig. 2. Evolution of particle positions with time for the case
without control and with feedback control given by Eq. (7'). Disk
diameter = 1 m; shell thickness = 0.025 m.

solutions of the plasma equations, experiments are performed on a
two-dimensional computer-simulated plasma to determine the
effect of some of the feedback controls on plasma confinement. The
zero-size-particle, nearest-grid-point method (Hockney, 1968)
(32 × 32 mesh) for charge position to density conversion is used
along with the fast Fourier transform method for potential computations. Results are obtained for a 1000-electron-ion-pair plasma
immersed in a static nonuniform magnetic field with z-component
$B_z(x) = 4B_{zo}/(5 - x)$ with $B_{zo} = 0.5$ webers/m^2 , $-1 \leq x \leq 1$, and
initially confined to a circular disk D with Maxwellian speed distributions truncated at 2.5 times the mean speed . The walls of the
square domain are absorbing. Figure 1 shows the number of ions
and electrons in D vs. time for the cases without control and with e,ℓ
feedback control given by Eqs. (7) and (7') with E^j_{max} = 5×10^5 v/m;
$G_j = 5 \times 10^5$ v-m for all j, and with $\alpha_i = 0.8$, $\alpha_e = 0.2$. The large
value of E^j_{max} and G_j is attributed to parameter scaling to achieve
equal order of magnitudes in the control and other forces appearing
in the particles' equations of motion. The outer shell D_s is a ring
with 30 equally sized cells and the outward normal $\overrightarrow{\eta}_j$ is directed
radially through the cell center. Thus, the feedback controls are
radial electric fields. As expected from the heavy weighting on the
ion outflow, the results (Fig. 1) indicate improvement in the ion
confinement and deterioration in electron confinement. The even-
tual particle leakage can be attributed to $\overrightarrow{\nabla}B^2$ drift and particle
absorption at the walls. Figure 2 shows a typical record of the
evolution of particle positions with time. The results reported here
are limited; effects of cell thickness and variations of the weighting
factors on plasma confinement, and the physical realization of
feedback controls have not been discussed.

ACKNOWLEDGMENT

    This work was supported by an AFOSR Grant No. AFOSR-68-
1547.

1.3 GEOMETRICAL LIMITATIONS IN PLASMA FEEDBACK SYSTEMS

Joseph M. Crowley
Department of Electrical Engineering
University of Illinois, Urbana, Illinois 61801

ABSTRACT

Based on the dispersion relation of the individual mode, a
stability criterion accounting for the geometrical effects which inher-
ently limit detection and enforcing processes of feedback systems
while neglecting the nonlinear mode coupling effects, is formulated.
If the criterion predicts instability, the plasma should be considered unstable. If the criterion predicts stability below a certain
value of feedback gain, however, the plasma may in fact be unstable
under those conditions; but a reduction in gain will always lead to
stability. The agreement between stable operating regions predicted
by this criterion and by earlier exact analysis shows acceptable
agreement for design purposes.

I. INTRODUCTION

Most analyses of plasma feedback systems make the assumption
of ideal continuum feedback which implies that the controlling force
at every point in the plasma can be made proportional to the disturbance from equilibrium at that point. Although this assumption
describes the fundamental effect of the feedback system, it cannot
predict the degradation of performance when the feedback is supplied
by some discrete external mechanism. One neglected effect is the
introduction of phase shifts caused either by reactive components
in the feedback loop or by finite propagation time in the plasma
when detection and enforcement occur at different points. Usually,
these phase shifts are considered by expressing small disturbances
of the plasma in some orthogonal expansion, and then considering
the effect of time delays on each of the normal modes. Unfortu-
nately, this procedure neglects feedback-induced mixing of the
originally orthogonal plasma modes which occurs in detector and
enforcer. The detector is sensitive to disturbances at a finite
number of points, which implies that much of the spatial information
of the mode must be lost in the detection process. In addition, the
enforcer excites many modes with various amplitude and phase
characteristics for any input signal, due to its finite size. Since
normal modes of the uncontrolled plasma are coupled to each other
through the feedback mechanism, the extent of the coupling increases
as the feedback gain increases.

These effects have been included in elastic and fluid systems (Melcher, 1965a, Melcher, 1966) by expanding all disturbances in the eigenmodes of the uncontrolled plasma. The eigenfrequencies in the presence of feedback control are then given by the solution of an infinite determinant. Although the predictions of this analysis agree quite well with the results of experiments, the method is somewhat complicated for use as a design tool. This paper suggests an instability criterion which includes the phase and amplitude effects imposed by geometry and electronics, while neglecting the coupling of modes.

II.   THE STABILITY CRITERION

If mode-coupling is neglected, each component in the plasma feedback system may be represented by a linear device whose transfer characteristics are determined as follows (Fig. 1):

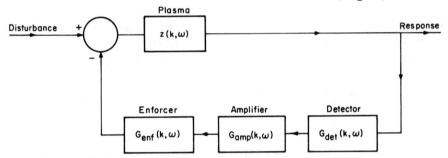

Fig. 1. Schematic Diagram of a Plasma Feedback System

1.   The response of the uncontrolled plasma $Z(k, \omega)$ to the control variable determined by means of a suitable linear model,

2.   The response of the detector $G_{det}(k, \omega)$, calculated by assuming a single normal mode of the uncontrolled plasma and determining the output of the detector.

3.   The gain of the amplifier $G_{amp}(k, \omega)$, similarly determined, and

4.   The gain of the enforcer, $G_{enf}(k, \omega)$, determined by assuming a single mode at its input and calculating magnitude and phase of the same mode produced in the plasma, while neglecting all other modes produced by the enforcer.

With this procedure, the feedback is assumed to operate on each mode individually so that the stability criterion (or dispersion relation) for that mode is given from standard feedback techniques as:

$$1 + GZ = 0 \qquad\qquad (1)$$

where
$$G = G_{det} G_{amp} G_{enf}$$

Before applying this criterion to a plasma feedback system, it is instructive to consider how its predictions may be affected by mode-coupling. Since the modes are coupled in the feedback loop, the coupling will have little effect when the loop gain is sufficiently small. As the gain increases, the effects of mode-coupling will become more important.

Although the effect of coupling must be determined for each system individually, it seems prudent to accept a prediction of instability at face value rather than hope that mode-coupling will improve the stability of a plasma. On the other hand, a prediction of stability should be treated with suspicion, particularly at high gain where mode-coupling is strong.

As an example of the reliability of the simple stability criterion, the elastic systems studied exactly by Melcher (1965a) were analyzed with the neglect of mode-coupling. The stability predictions of both approaches for one of those systems are shown in Fig. 2 depicting

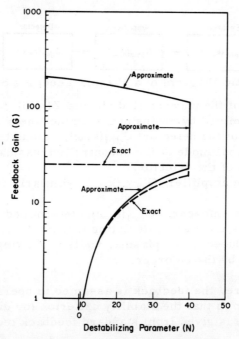

Fig. 2.   Comparison of the Coupled-Mode Stability Theory
with the Present Approximation

lines of neutral stability for various combinations of the destabilizing
parameter N and feedback gain G. As expected, the two criteria
agree very well at low values of G, but show some divergence at
larger gains.

III. APPLICATION

In the first successful plasma feedback experiments, six feedback stations at the circumference of the OGRA machine controlled
a flute mode (Arsenin and Chuyanov, 1968). In this case, the plasma
response function was

$$Z = \frac{(a/b)^m (\omega^2 - m\omega)}{\omega^2 - m\omega + |m| \beta_m N} \tag{2}$$

where β_m is a geometrical constant. Detector, amplifier, and
enforcer gains may be written as

$$G_{det} = 1, \tag{3}$$

$$G_{amp} = G \text{ and} \tag{4}$$

$$G_{enf} = \frac{\sin(m\pi/6)}{m\pi} \tag{5}$$

Application of the stability criterion results in a stability plot
(Fig. 3) which shows neutral stability for various modes as a function of normalized feedback gain and density.

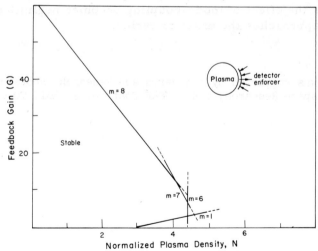

Fig. 3. The Approximate Stability Disgram for the
OGRA Feedback Experiment

This plot is interpreted as follows: The m = 1 mode becomes unstable as the density (N) is increased. With no feedback (G = 0), instability occurs at N = 2.91, but the critical density may be raised by increasing the feedback gain (G). Modes m = 2, 3, 4, 5 behave similarly, but the m = 6 mode, which cannot be excited by the enforcer, is unstable above a density N = 4.42 for all values of gain. This represents the performance limit of the feedback system. Higher modes, such as m = 7, 8, experience phase shifts in the enforcer which drive these modes unstable when the gain is high. Although Fig. 3 represents the stability diagram for a plasma, it is quite similar to that of an elastic system (Fig. 2) because the geometry of the detector and enforcer is similar in both cases.

Thus, the simple criterion indicates that the feedback system allows attainment of higher densities by stabilizing the first five modes, but that the sixth mode will become unstable when the density has increased by a factor of 1.52. This is in approximate agreement with the results of the experiment in which the plasma remained stable until the density was approximately doubled, at which point an (unspecified) high-frequency instability appeared.

IV. SUMMARY

An approximate stability criterion is formulated for plasma feedback systems by neglecting mode-coupling in the detector and enforcer. This criterion should be interpreted as a conservative estimate of the unstable operating regions. At sufficiently low feedback gains the effect of mode-coupling becomes negligible and the criterion approaches the exact criterion.

ACKNOWLEDGMENT

This work was supported by the Aerospace Research Laboratories, Office of Aerospace Research, under USAF Contract F 33615-70-C-1091.

1.4 GENERALIZED BOUNDARY CONDITIONS OF PLASMA FEEDBACK SYSTEMS

E. L. Lindman
Department of Physics, University of Texas
Austin, Texas 78712

ABSTRACT

General boundary conditions of plasma-feedback systems using electrodes as both sensor and suppressor at plasma boundaries simultaneously are discussed. Difference equations describing the electrical properties of the boundary are derived and converted to differential equations which lead to a boundary dispersion relation describing the boundary modes and their interaction with the plasma. Application to Kelvin-Helmholtz instability is given.

I. INTRODUCTION

The concept of generalized boundary conditions (Lindman, 1967), employing electrodes at plasma boundaries as both sensing and suppressing elements simultaneously, arises when considering the application of feedback techniques to the stabilization of thermonuclear plasmas which have a large number of degrees of freedom and where the type of motion to be stabilized is wave-like in character. A system capable of interacting with a set of waves is a set of sensing and control electrodes whose spacing is small compared to the shortest wavelength. In other words the number of degrees of freedom of the system must exceed the number of degrees of freedom represented by the set of waves to be stabilized. This large number of closely spaced sensing and control electrodes leads to difficulties in isolation. A strong control voltage applied to the control electrode will affect the neighboring sensing electrodes. If such a system is itself not unstable the control electrodes will severely limit the sensitivity of the sensing electrodes.

Consequently one is all but forced to consider systems in which the electrodes are employed simultaneously as both sensing and control electrodes. At this point we leave conventional feedback control theory and arrive at a control system which consists of a large number of electrodes connected by active or passive circuitry which surrounds the plasma. Since it forms a boundary to the plasma, its effect on the plasma is most conveniently described in terms of a boundary condition.

There are further constraints on the system, however, which arise because of the nature of thermonuclear plasmas. The first arises from the toroidal nature of the plasma while the second arises from the occurrence of very strong electromagnetic fields in the case of a pulsed device. The first constraint requires that the control system act across B and that the "ends" of the plasma not be accessible. The second leads to great difficulty in the use of active circuitry, thus focussing attention on the use of passive circuitry.

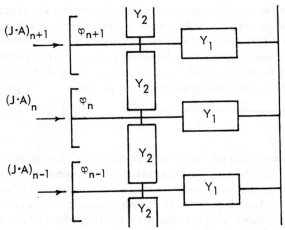

Fig. 1. A section of the segmented boundary and the associated electrical network is shown. By varying the admittances Y_1 and Y_2 the properties of the plasma-boundary modes can be changed and stabilization obtained. The actual boundary condition arises from the fact that the current flowing from the plasma to the boundary segment must equal the current flowing into the network from the segment and the segment potential must be equal to the local plasma potential. (Sheath effects have been neglected.)

These considerations lead to the control system shown in Fig. 1, which is essentially a "lumped-parameter" transmission line in electrical contact with the plasma. Passive admittances, (Y_1's and Y_2's) only, are considered. The boundary condition is obtained from the requirements :

1) the boundary segments be at the local plasma potential and
2) the plasma current flowing to the boundary segment be balanced by the current flowing into the network.

Circuit analysis leads to a difference equation, which is converted for mathematical convenience (assuming the segment dimensions are small) into a differential equation. This differential equation then leads to a boundary dispersion relation which describes the boundary modes and their interaction with the plasma. Solving the plasma dispersion relation for the waves of interest together with this dispersion relation gives the properties of these waves. A more detailed description of generalized boundary conditions and a demonstration of their stabilizing effect on universal ion-sound and drift-wave instabilities is given elsewhere.[*]

II. APPLICATION TO EDGE OSCILLATIONS

A test of this approach to plasma stabilization is being carried out on the Q-machine at the University of Texas. Edge oscillations are observed and a stabilizing boundary in which $Y_2=0$ and $Y_1=i\omega C$ is applied.

For the case in which $Y_2=0$ the segments decouple. For each segment Ohm's Law gives

$$AJ_r = i = Y_1\phi \tag{1}$$

where A is the area of the segment, J_r is the plasma current density flowing into the segment, i is the current flowing in the admittance Y_1, and ϕ is the potential drop across Y_1. By neglecting finite-Larmor-radius and sheath effects the current flowing to the segment is the ion dielectric current

$$J_r = \epsilon_o \frac{\omega_{pi}^2}{\Omega_i^2} \left[-i(\omega + m\omega_2) \frac{\partial\phi}{\partial r} \right]_{r=r_b} \tag{2}$$

where ϵ_o = capacitivity of free space, ω_{pi} is the ion plasma frequency, Ω_i is the ion gyro frequency, ω is the wave frequency, ω_2 is the plasma rotation frequency in the outer region, and $-\partial\phi/\partial r$ is the radial electric field.

The description of the plasma waves is given by Eq. (8) of Kent, Jen, and Chen (1969):

---

[*] E. L. Lindman, to be published in Physics of Fluids.

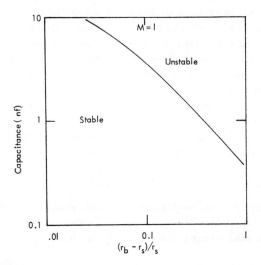

Fig. 2.   The boundary shown in Fig. 1 is applied to edge oscillations.   Approximate dispersion relations are obtained from the
two and three region approximations given by Kent, Jen, and Chen
(1969).   The finite- Larmor radius terms were set to zero and typical
Q-machine parameters were chosen: $N_o = 3 \times 10^9 cm^{-3}$, B=1.5 k
Gauss, r_s=1.3 cm.   For M=1, stability is obtained for boundary cap-
acitances below a critical value which depends on the radius to the
discontinuity, r_s, and the radius to the boundary r_b.   For this case
Y_2=0 and $Y_1=i\omega C$.

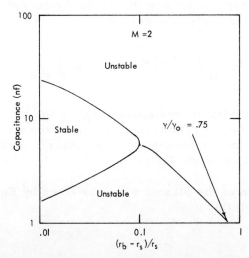

Fig. 3.   Stability region for the M=2 case.   The lower
unstable region has a reduced growth rate which varies in magnitude.
At r_b=1.77 r_s a reduction of .75 is obtained.   See Fig. 2 for further
data.

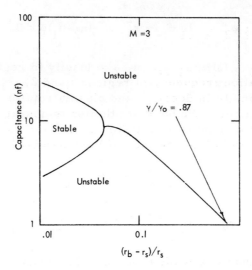

Fig. 4. Stability region for the M=3 case. The lower
unstable region has a reduced growth rate which varies in magnitude. At $r_b = 1.77\ r_s$ a reduction of .87 is obtained. See Fig. 2 for
further data.

$$T_2 \phi_2' - T_1 \phi_1' + r_s^2 \omega^2 (n_2 - n_1)\phi = 0 \tag{3}$$

where $T_{1,2} = r^3 n_{1,2} (\omega + m\omega_{1,2})$, $n_{1,2}$ = density in regions 1 and 2, $\omega_{1,2}$ = plasma rotation frequency in regions 1 and 2, r_s = radial location of discontinuity in density and plasma rotation, and m = azimuthal wave number. We have set the ion temperature equal to zero in order to avoid confusion with finite- Larmor- radius stabilizing effects. The radial dependence of the potential in the two regions is taken as

$$\phi_1 = A\, r^{-1 + |m|}; \quad 0 \le r \le r_s, \tag{4}$$

$$\phi_2 = B\, r^{-1 + |m|} + C\, r^{-1 - |m|}; \quad r_s \le r \le r_b. \tag{5}$$

Combining these equations leads to a 4th order equation for ω which was evaluated numerically.

The following general results are obtained:
1). The stabilizing effect of the boundary decreases as the boundary is moved away from the region in the plasma at which the wave is localized; and 2). For fixed boundary, the stabilizing effect decreases as the transverse wave numbers (e.g. as m and k_z) increase in a cylindrical system.

For the M = 1 mode, stabilization is obtained when the boundary capacity is reduced below a critical value. The dependence of this critical capacity on the boundary radius and radius to the destabilizing sheath is shown in Fig. 2.

For the M = 2 and M = 3 modes the situation is more complicated. For small boundary radius, stability is obtained in between two values of the capacity. At larger radii complete stabilization disappears as shown in Figs. 3 and 4. However reduced growth rates are obtained in the lower unstable region. The reduction depends on the radius and typical reductions at $r_b = 1.8 r_s$ are .75 for the M = 2 case and .87 for the M = 3 case. As r_b approaches r_s larger reductions in growth rate are obtained, of course.

Experimental results (Carlyle, 1970) are in reasonable agreement with these calculations. Further studies, both experimental and theoretical, of more complicated networks are being carried out.

1.5 PLASMA STABILIZATION BY FEEDBACK

J. B. Taylor and C. N. Lashmore-Davies

U.K.A.E.A., Culham Laboratory, Abingdon, Berkshire, England

ABSTRACT

A simple model is discussed which illustrates the general features of plasma stabilization by an external feedback system. This indicates that different phase relations in the feedback loop are needed to stabilize differing classes of electrostatic instability.

I. INTRODUCTION

Any practical application of the method of feedback control to stabilize a plasma depends on the details of the particular instability and of the apparatus. However, a simple model can account for most of the qualitative features of any particular system and also illustrates some significant differences in the application of feed-back stabilization to different types of electrostatic instability.

In this model the feedback mechanism is one which senses the potential at r' and in response to this signal charges up a suppressor element at r. The response function of the feedback circuit may be written as:

$$g(\omega)\, G(r, r')$$

where $G(r, r')$ is real and symmetric[†] and $g(\omega)$ complex. The modulus of $g(\omega)$ specifies the amplification and the argument of $g(\omega)$, the phase difference introduced by the feedback circuit. Since the feedback must be real and causal, $g(\omega) = g^*(-\omega)$ and $g(\omega) \to$ real constant as $\omega \to \infty$; it must also satisfy the Kramers-Kronig dispersion relation (Landau and Lifshitz, 1960a).

In the absence of feedback the response of the plasma to an electric potential $\phi(r)e^{-i\omega t}$ can be represented by a generalized conductivity tensor $\kappa_\omega(r, r')$ such that:

$$J_\omega(r) = \int \kappa_\omega(r, r')\, E_\omega(r')\, dr'. \qquad (1)$$

This leads to the dispersion equation:

[†]This condition does not affect the final results but serves only to simplify the algebra.

$$\nabla \cdot \int dr' \ \varepsilon_\omega(r, r') \ \nabla \phi(r') \ = \ 0, \tag{2}$$

where

$$\varepsilon_\omega(r, r') \ = \ \delta(r, r') \ - \ \frac{4\pi}{i\omega} \kappa_\omega(r, r'). \tag{3}$$

Equation (2) determines the oscillation frequencies and the stability of the system through the eigenvalue ω.

When the suppressor is in operation it introduces 'external' charge given by:

$$\rho_{ext}(r) = \ g(\omega) \int dr' \ G(r, r') \ \phi(r'), \tag{4}$$

so that with the application of feedback the oscillation frequencies are given by:

$$\nabla \cdot \int dr' \ \varepsilon_\omega(r, r') \ \nabla \phi(r') + g(\omega) \int dr' G(r, r') \phi(r') = 0. \tag{5}$$

Before the effect of the feedback can be discussed it is important to distinguish two different types of instability which can arise from Eq. (2). The first type involves only a single mode of oscillation, which may have positive or negative energy. Growth of this instability is accompanied by an exchange of energy between the oscillation and the plasma medium, e.g. by dissipation. The second type of instability involves two modes of oscillation, one of positive and the other of negative energy. These oscillations become degenerate at the threshold of instability, which can be regarded as the exchange of energy between the two oscillations without any net transfer to the plasma medium. Following Hasegawa (1968) we refer to the first instability as dissipative and the second as reactive and we consider the two cases separately.

II. DISSIPATIVE INSTABILITIES

If the plasma were neutrally stable it would neither absorb nor emit energy when acted on by an electric field and the power absorbed would be zero. The necessary and sufficient condition for this (Drummond, Gerwin, and Springer, 1961) is that $\varepsilon(r, r')$ be hermitian, that is:

$$\varepsilon_{ij}(r, r') \ = \ \varepsilon_{ji}^*(r, r') \tag{6}$$

Thus, if we write:

$$\varepsilon \ = \ \varepsilon_h + \varepsilon_a \tag{7}$$

where ε_h and ε_a are the hermitian and anti-hermitian parts of ε respectively, then if the plasma is only weakly unstable, ε_a will be small compared to ε_h. Using standard perturbation theory, treating both ε_a and the suppressor term $g(\omega)$ as small compared to ε_h, we can determine the effect of the feedback on the real and imaginary parts of $\omega = \omega_o + i\gamma_o$ as the following:

$$\omega = \omega_o + \kappa g \cos \theta, \tag{8}$$

$$\gamma = \gamma_o + \kappa g \sin \theta, \tag{9}$$

where

$$g \equiv |g(\omega_o)| \; ; \; \theta \equiv \arg g(\omega_o),$$

and

$$\kappa = \frac{\iint \phi^*(r) \, G(r, r') \, \phi(r') \, dr \, dr'}{\iint \nabla \phi^*(r) \, \frac{\partial}{\partial \omega} \varepsilon_h(r, r') \bigg|_{\omega = \omega_o} \nabla \phi(r') \, dr \, dr'} \tag{10}$$

where κ is real due to the properties of $\varepsilon_h(r, r')$ and $G(r, r')$. Equations (8) and (9) show that the effect of the feedback may be to suppress or enhance instabilities according to the phase difference θ which is introduced between sensor and suppressor elements; a stable system may be rendered unstable and vice versa. The condition for stabilization is:

$$-\Im m \, g(\omega_o) = -g \sin \theta > \gamma_o/\kappa, \tag{11}$$

that is, the amplification must be greater than a critical value and the phase must be in one half of the phase plane for a positive energy wave and the opposite half for a negative energy wave. Furthermore, as a function of phase angle θ the effect of feedback on the real frequency is zero when the instability suppression or enhancement has its greatest value. These general features are exhibited in the experiments of Simonen, Chu, and Hendel (1969) and of Keen (1970), where the fluctuation level serves as a measure of γ_o to which it is related through nonlinear limiting effects.

III. REACTIVE INSTABILITIES

The reactive instability does not depend on dissipation and we may neglect ε_a. At the threshold of stability we have:

$$\iint \nabla \phi^*(r) \frac{\partial}{\partial \omega} \varepsilon_h(r, r') \, \nabla \phi(r') \, dr \, dr' = 0. \tag{12}$$

This condition represents the fact that the threshold point is reached as a positive and negative energy mode become degenerate. To apply perturbation theory to this case we treat both the deviation from threshold and the suppressor term as small quantities, i.e. we expand about the marginally stable state. Then, making use of the above condition one obtains for the perturbed eigenvalue:

$$(\omega - \omega_o)^2 = -\gamma_o^2 + g(\omega)\kappa \tag{13}$$

where now:

$$\kappa = \frac{2 \iint \phi^*(r)\ G(r, r')\ \phi(r')\ dr\ dr'}{\iint \nabla \phi^*(r)\ \left.\frac{\partial^2}{\partial \omega^2}\varepsilon_h(r, r')\right|_{\omega=\omega_o}\nabla \phi(r')\ dr\ dr'}, \tag{14}$$

ω_o is the oscillation frequency at threshold and γ_o the growth rate without feedback.

From Eq. (13) the conditions for stability are:

$$\mathcal{R}e\ g(\omega_o) > \gamma_o^2/\kappa, \tag{15}$$

$$\mathcal{I}m\ g(\omega_o) = 0 \tag{16}$$

$$\mathcal{I}m\ g'(\omega_o) > 0 \tag{17}$$

In other words, we again have the result that the amplification must be greater than some critical value but this time any phase difference other than zero produces instability. Evidently this is due to the presence of a positive and negative energy wave at the same value of frequency. Notice that if the instability is a purely growing mode ($\omega_o = 0$) equation (16) is automatically satisfied.

IV. CONCLUSIONS

We have shown that for a dissipative instability one half of the phase plane can produce stability whereas for the reactive case only one value of the phase is stabilizing. For $\omega_o = 0$ most circuits will automatically satisfy this condition therefore the $\omega_o \neq 0$ case is the most serious.

1.6 STABILIZATION OF A LOW-DENSITY PLASMA IN A SIMPLE MAGNETIC MIRROR BY FEEDBACK CONTROL

C. N. Lashmore-Davies
UKAEA Research Group, Culham Laboratory,
Abingdon, Berkshire, England

ABSTRACT

Stabilization of an electrostatic flute-type instability occurring in a simple magnetic mirror by feedback techniques is discussed. In the first part of the paper a diffuse plasma is considered. The effect of varying the locations of the sensing and suppressing systems is found to alter the stability threshold significantly. In the second part a sharp-boundary plasma is considered and phase shift and frequency response are included in the feedback terms.

I. INTRODUCTION

The possibility of stabilizing the various instabilities which occur in a hot confined plasma by feedback techniques has recently received a good deal of attention. Electrostatic instabilities can be divided into two general types, dissipative and reactive (Hasegawa, 1968). The dissipative instability is produced by one wave (whose energy can have either sign) being driven unstable due to a net flow of energy between the oscillation and the plasma medium. The reactive instability is due to two waves whose energies are of opposite sign becoming degenerate in their frequency without any net energy flow between oscillation and medium. This case can be thought of as an exchange of energy between two modes of oscillation.

The conditions for stabilizing the two types of instability are very different, at least at threshold (Taylor and Lashmore-Davies, 1970). Most experiments (and theories) on plasma stabilization by feedback have dealt with dissipative instabilities. An exception to this is provided by the work of Arsenin and Chuyanov (1968) where the problem of stabilizing a simple magnetic mirror against a flute type instability was considered.

In this paper we extend the model of Arsenin and Chuyanov (1968) to include the effects of a diffuse plasma and different locations of sensing and suppressing systems. The occurrence of nodes in the oscillation amplitudes of a bounded plasma may be expected to have a significant effect on any feedback system. Finally for a

sharp-boundary plasma we take account of phase shift and frequency
dependence of the feedback system to show that the stabilization
conditions of a reactive instability may not be quite so stringent as
the threshold calculation implies (Taylor and Lashmore-Davies,
1970).

II. STABILIZATION CONDITIONS FOR A DIFFUSE PLASMA

We assume a cylindrical plasma of infinite length whose z-axis
coincides with the direction of a uniform constant magnetic field.
The plasma is nonuniform of radius a and we simulate magnetic
curvature and gradients with a radial gravitational force. Assuming
the perturbations are electrostatic and vary as:

$$\phi(\underline{r}, t) \propto \phi(r) \exp\left[i(m\phi - \omega t)\right],$$

we obtain the equation (Arsenin and Chuyanov, 1968):

$$\left(1 + \frac{\omega_{pi}^2}{\Omega_i^2} N(x)\right) \frac{d^2\phi}{dx^2} + \left[\left(1 + \frac{\omega_{pi}^2}{\Omega_i^2} N(x)\right) \frac{1}{x} + \frac{\omega_{pi}^2}{\Omega_i^2} \frac{dN}{dx}\right] \frac{d\phi}{dx}$$

$$- \frac{m^2}{x^2}\left(1 + \frac{\omega_{pi}^2}{\Omega_i^2} N(x)\right)\phi + m^2 \frac{\omega_{pi}^2}{\Omega_i} \frac{\omega^*}{\omega(\omega + m\omega^*)} \frac{1}{x} \frac{dN}{dx}\phi = 0,$$

$$(1)$$

where x is r/a, N(x) gives the density profile and ω^* is the
precession frequency of the ions due to the gravitational drift
($\omega^* = g/\Omega_i a$) where the radial gravitational force was taken as
g(r) = gr/a. The remaining quantities have their usual meaning and
ω_{pi} refers to the cylinder axis. For a parabolic profile:

$$N(x) = 1 - x^2$$

Eq. (1) reduces to Bessel's equation provided $\omega_{pi}^2 \ll \Omega_i^2$.

The feedback condition is obtained by introducing a surface
charge density $\sigma(x_2)$ at the surface $x = x_2$ (the suppressor)

$$\sigma(x_2) = \frac{\delta}{a} \epsilon_0 \phi(x_1) \qquad (2)$$

where $x = x_1$ is the sensing surface and δ is real and represents
the amplification of the feedback circuit (without phase shift). The
boundary conditions at $x = x_2$ are:

$$\phi^{I}(x_2) = \phi^{II}(x_2) \tag{3}$$

and

$$\frac{d}{dx}\phi^{II}(x_2) - \frac{d}{dx}\phi^{I}(x_2) = \delta\,\phi(x_1) \tag{4}$$

where ϕ is the solution at the sensor, ϕ^{I} is the solution for $x < x_2$ and ϕ^{II} is the solution for $x > x_2$. The remaining boundary conditions are the continuity of ϕ and $d\phi/dx$ at $x = 1$ and $\phi \to 0$ as $x \to \infty$.

(i) <u>Sensor and Suppressor Outside the Plasma</u>

For $x_1 = 1$, $x_2 > 1$, and $m = 1$ (this example turns out to be independent of x_2) applying the above boundary conditions gives the following dispersion relation:

$$p\,J_0(p)/J_1(p) = -\delta \quad, \tag{5}$$

where

$$p^2 = -2m^2\,\omega_{pi}^2\,\omega^*/\Omega_i\,\omega(\omega + m\omega^*) \quad. \tag{6}$$

Comparing the stability thresholds with and without ($\delta = 0$) feedback with the aid of Fig. 1 we see that a large amount of positive ($\delta \gg 1$)

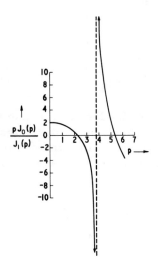

Figure 1

feedback[†] improves the density threshold by a factor 2 whereas there is an optimum value ($\delta \leq -2$) of negative feedback which increases the density by a factor 4.

(ii) <u>Sensor Inside the Plasma</u>

For this case $x_1 = \frac{1}{2}$ and the dispersion relation for m = 1 is:

$$p\, J_0(p)/J_1(p/2) = -\,\delta \qquad\qquad (7)$$

which is again independent of x_2. An examination of Fig. 2 shows

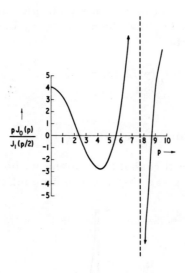

Fig. 2

that both positive and negative feedback increase the threshold (i.e. the larger the value of p at which the root occurs the larger the density threshold). However, this time the optimum value of amplification occurs for positive feedback increasing the density threshold by a factor of 12.

(iii) <u>Sensor and Suppressor Inside the Plasma</u>

We again treat the m = 1 case and take $x_1 = \frac{1}{2}$, $x_2 = \frac{3}{4}$ giving the dispersion equation:

[†]In this paper positive feedback refers to $\delta > 0$ and negative feedback to $\delta < 0$.

$$\frac{J_0(p)}{J_1(p/2)} \left\{ Y_1(3p/4) J_0(p) - J_1(3p/4)Y_0(p) \right\}^{-1} = \frac{3\pi}{8} \delta \quad . \tag{8}$$

Obtaining the roots of this equation graphically (from Fig. 3)

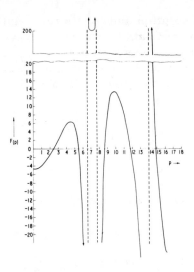

Fig. 3

we again find that both positive and negative feedback produce an improvement in the density threshold and that positive feedback is more effective. The density is increased by a factor of 36 by the optimum positive feedback. Note that if δ is made too large the density threshold is reduced by a factor 4, i.e.

$$14 \leq \delta < 200 \quad . \tag{9}$$

(iv) Frequency Dependence of the Suppressor

In the previous examples the phase shift was always 0 or π. However, it was shown (Taylor and Lashmore-Davies, 1970) that at threshold the slightest departure from this condition resulted in instability. Here we consider the stability of a sharp boundary plasma,

$$N = 1 \quad \text{for} \quad x < 1$$
$$N = 0 \quad \text{for} \quad x > 1 \quad ,$$

but allow for a specific form of frequency dependence. The boundary conditions are the same as before except that at $x = 1$:

$$\frac{d}{dx}\,\phi_v^I(1) - \frac{d}{dx}\,\phi_p(1) - \frac{\omega_{pi}^2}{\Omega_i^2}N(1)\,\frac{d}{dx}\,\phi_p(1) = m^2\,\frac{\omega_{pi}^2}{\Omega_i}\,\frac{\omega^*}{\omega(\omega+m\omega^*)}\,N(1)\phi_p(1)$$

(10)

where ϕ_p is the plasma solution and ϕ_v^I is the solution from
$x = 1$ to $x = x_2$. This corresponds to a surface-wave problem and
can be solved for arbitrary values of the ratio ω_{pi}/Ω_i. Applying the
stated boundary conditions with both sensing and suppressing systems
outside the plasma we obtain the dispersion relation for $m = 1$:

$$2 + \frac{\omega_{pi}^2}{\Omega_i^2} + \frac{\omega_{pi}^2}{\Omega_i}\,\frac{\omega^*}{\omega(\omega + \omega^*)} + \delta(\omega) = 0 \ .$$

(11)

Taking an integral feedback condition

$$\delta(\omega) = \alpha + i\eta/\omega$$

(12)

we obtain the following conditions for stability from a Nyquist
analysis:

$$\eta/\omega^* < 0 \quad ; \quad -\alpha > 2 + \omega_{pi}^2/\Omega_i^2 \ .$$

(13)

In other words above a certain level of amplification one quarter of
the phase plane produces stability.

III. CONCLUSIONS

It has been shown that for a diffuse plasma there is an optimum
value of the amplification for stability. If the sensor is outside the
plasma negative feedback is required. If it is inside, the opposite
holds. The physically reasonable result was obtained that the
closer the feedback system to the plasma the greater its effectiveness. Finally, a time integral feedback was shown to allow stability
over one quarter of the phase plane.

1.7 PLASMA CONTROL WITH INFRARED LASERS

F. F. Chen

University of California, Los Angeles 90024

ABSTRACT

An extraordinary electromagnetic wave of large amplitude will produce a quasilinear dc drift of electrons relative to ions, which can be modulated for feedback stabilization of low-frequency waves. The effect occurs at the upper hybrid resonance. A promising way to penetrate the cut-off in a fusion plasma is to use the nonlinear interaction of two CO_2 laser beams to produce a difference frequency near ω_h. We have extended the cold plasma analysis for this process and find that the intensities required are not unreasonable even when thermal effects are taken into account.

I. INTRODUCTION

The central problem in feedback stabilization of fusion reactor plasmas is the introduction of the feedback signal into the interior of a dense plasma. The only scheme proposed so far (Chen and Furth, 1969) requires the generation of intense beams of high-energy neutral atoms, a rather cumbersome affair. In this paper, we suggest the use of CO_2 lasers, which are becoming powerful enough to be considered for plasma production and heating and are therefore surely powerful enough for plasma control.

We take advantage of a nonlinear interaction between an extraordinary electromagnetic wave ($\underset{\sim}{E}$, $\underset{\sim}{k} \perp B_0$) and a plasma to produce a dc drift of electrons relative to ions. This second-order drift, (Chen and Etievant, 1970) for a cold plasma is:

$$\underset{\sim}{v}_e^{(2)} = -4\pi \frac{\underset{\sim}{P}}{B_o^2} \frac{\omega_c^4}{\omega_p^4} \frac{1}{\delta^2}, \qquad \delta \equiv \frac{\omega_h^2 - \omega^2}{\omega_p^2}, \qquad (1)$$

where ω_h is the upper hybrid frequency, B_o is in gauss, and $P = (c/8\pi)(ck/\omega)E_t^2$ is the Poynting vector in ergs per cm^2 per sec, E_t being the transverse component of the elliptically polarized X-wave E-vector. Since k varies as $\delta^{-1/2}$ near hybrid resonance, $\underset{\sim}{v}_e^{(2)}$ varies as $\delta^{-5/2}$; and the effect is localized to the density layer where $\omega \approx \omega_h$.

If $v_e^{(2)}$ is directed along ∇n_0, it will cause a charge separation and an $\tilde{E}_1$. By feedback-modulating $v_e^{(2)}$, it is possible to suppress low-frequency waves. Results for resistive drift waves are given in Sec. II. That a nonlinear effect of the this type actually occurs may have been demonstrated by Wong, Baker, and Booth (1970) whose experimental results inspired this work.

Two major difficulties arise in applying this method to a fusion plasma. First, the free-space wavelength λ corresponding to hybrid resonance in a plasma with $n = 10^{15} cm^{-3}$ and $B_0 = 10^5 G$ is 750μ, much longer than the 10.6μ wavelength of CO_2 lasers. Second, a wave launched outside the plasma will be reflected at the right-hand cutoff

$$\omega_R = \frac{1}{2} [\omega_c + (\omega_c^2 + 4\omega_p^2)^{1/2}]$$

and will never reach hybrid resonance. Tunneling through the cutoff was possible in a microwave experiment (Wong, Baker, and Booth, 1970) where $\lambda >> r_0 \equiv (-n_0/n_0')$. The opposite is true in a fusion reactor, and computations by Kuehl (1967) show that tunneling is then negligible. However, both difficulties can be overcome by the optical mixing of two CO_2 or N_2O laser beams to produce a difference frequency near ω_h. This process is treated in Sec. III.

II. STABILIZATION OF DRIFT AND FLUTE MODES

Following the procedure and notation of Chen and Furth (1969), we consider $v_e^{(2)}$ to have the same $\exp[i(k_y y + k_z z - \omega t)]$ dependence as the drift-wave variables and add it to the usual solution of the electron equation of motion. The usual resistive drift wave analysis is then carried out for $B_0 = B\hat{z}$ and $\nabla n_0/n_0 = \hat{x}/r_0$. Gravitational instabilities are included via a term $M n_0 g\hat{x}$ in the ion equation of motion. We define

$$\omega_f \equiv [v_{ex}^{(2)}/r_0 + i v_{ey}^{(2)} k_y] (n_0/n_1).$$ (2)

For g=0, there results the local dispersion relation

$$\omega(\omega - \omega_i) - i\omega_f(b^{-1}\omega^* - \omega + \omega_i) + i\omega_s[\omega - \omega^* + b(\omega - \omega_i)] = 0$$ (3)

This has been solved numerically to obtain curves showing the optimum phase of $v_e^{(2)}$ relative to n_1. For $\omega \approx \omega^*$, sufficient conditions for marginal stability are

$$v_{ex}^{(2)} = 2ib\omega^* r_0 (n_1/n_0) \text{ and } v_{ey}^{(2)} = 2b(\omega^*/k_y)(n_1/n_0)$$ (4)

for "radial" and "tangential" beams, respectively. These values are independent of ω_s (resistivity) because $v_e^{(2)}$ cancels directly the

charge separations caused by ion inertia and FLR, and parallel
electron currents need not flow. We believe, therefore, that Eq. (4)
has rather general applicability.

For $g \neq 0$ and $k_z = 0$, the dispersion relation is

$$(\omega + i\omega_f)(\omega - \omega_i + k v_g) - g/r_o - i\omega_f \omega^* b^{-1} = 0, \tag{5}$$

where $v_g = -g/\Omega_c$. For radial injection, gravitational flute modes
are stabilized by $v_{ex}^{(2)} = (ibg/\omega^*)(n_1/n_o)$.

III. NONLINEAR CONVERSION TO $\omega \approx \omega_h$.

We consider two ordinary waves (E $\parallel B_o$) $\omega_1 >> \omega_h$ and $\omega_2 >> \omega_h$
propagating at a relative angle θ_{12} in the plane perpendicular to B_o.
If we make $\omega_1 - \omega_2 = \omega_3 \approx \omega_h$, nonlinear mixing will form an extraordinary wave propagating at an angle θ_{13} relative to ω_1, also in the
plane perpendicular to B_o. The wave ω_3 can then produce the drift
$v_e^{(2)}$. It is essential that the polarizations of ω_1, ω_2, and ω_3 be as
stated above. Since the result will depend on the fourth power of the
incident-wave amplitudes, the estimates given below, based on
weakly nonlinear theories, are probably pessimistic.

The conversion of ω_1, ω_2 to ω_3 has been computed by Etievant,
Ossakow, Ozizmir, Su, and Fidone (1968) for a cold plasma. The
result can be written as:

$$P_3 = 5.7 \times 10^{-18} \pi R_o^2 P_1 P_2 F(\omega_1, \omega_2, \omega_c, \omega_p), \tag{6}$$

where πR_o^2 is the area of intersection of the incident beams. For
$\omega_3 \approx \omega_h$, the linear dispersion relations give $k_3^2 \approx \omega_c^2/c^2\delta$ and
$k_{1,2}^2 = \omega_{1,2}^2/c^2$, whence F takes the form

$$F \approx \frac{\omega_3 \omega_c^3}{\omega_1^2 \omega_2^2} \delta^{-5/2}. \tag{7}$$

Computations of log F, $\cos \theta_{12}$, and $\cos \theta_{13}$ are shown in Fig. 1.
Our computations differ from those of Etievant et al. in that the wave
ω_3 is not required to propagate out of the plasma; it may be trapped
and converted to $v_e^{(2)}$ locally. The restriction $\omega_p \lesssim \omega_c$ is then
removed. F is sharply peaked near $\delta = (\omega_h^2 - \omega_3^2)/\omega_p^2 = 0$. In
cold-plasma theory, F has a maximum value F_{max} imposed by the
condition $k_3 = k_1 - k_2$, which causes $\cos \theta_{12} < -1$ and $\cos \theta_{13} > 1$
for $\delta < \delta_{min}$. Thus, an upper limit to P_3 is given by Eq. (6) and

$$F = F_{max} \approx 32 \omega_1 \omega_h/\omega_c^2. \tag{8}$$

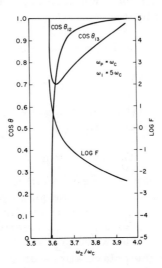

Fig. 1

Note that setting $\theta_{12} = 180°$ automatically ensures operation at F_{max}.

From Eqs. (1), (6), and (7), one finds that $v_e^{(2)}$ varies as $\delta^{-9/2}$. Clearly, a very accurate computation of the width of the hybrid resonance is required. In the absence of such a calculation, we shall give an estimate of the thermal corrections. First, we have extended the work of Chen and Etievant (1970) and Etievant et al. (1968) to include a ∇p term; this procedure splits the singularity at $\delta = 0$ but does not remove it. Second, we have used the condition $\omega/k \gtrsim v_{th}$, where $v_{th}^2 \equiv 3KT_e/m$. This yields a result intermediate between our optimistic estimate, Eq. (8), and our pessimistic estimate. Finally, we have made a pessimistic estimate on the basis that finite electron Larmor radius (FLR) effects broaden the resonance. Hedrick (1970) has shown that k for X-waves with FLR becomes complex near $\omega = \omega_h$, where the X-wave couples to the Bernstein modes. If we require k to be real, we obtain the condition $\delta \gtrsim v_{th}/c$. Using this in Eqs. (1) and (7), we obtain a pessimistic estimate because nonlinear FLR effects have been neglected.

Estimates have been made for a plasma with $n=10^{15}cm^{-3}$, $B=10^5G$, $T_e = 10$ kev, $R = 2r_0 = 50$ cm, $n_1/n_0 = 10^{-3}$, and an azimuthal mode number m.   A single 10.6μ beam operating off-resonance would require $P=2\times10^7 m^3 W/cm^2$, according to Eqs. (1) and (4a). A single beam with $\omega \approx \omega_h$, $\delta = v_{th}/c$, would require only $P=1.3\times10^{-2}m^3 W/cm^2$, but the beam would not penetrate. Two CO_2 laser beams with $\pi R_o^2 = 1$ cm^2 would require $P_1 = P_2 = 2.4\times10^{-2}m^{3/2}W/cm^2$ optimistically and $P_1 = P_2 = 2\times10^7 m^{3/2}$ W/cm^2 pessimistically. We conjecture that the latter is a gross overestimate.

We have benefited from discussions with Profs. B. D. Fried and A. Y. Wong and from the UCLA computing facilities.

FEEDBACK THEORIES FOR MHD MODES

2.1 FEEDBACK STABILIZATION OF HYDROMAGNETIC CONTINUA:
REVIEW AND PROSPECTS

J. R. Melcher
Department of Electrical Engineering
Massachusetts Institute of Technology
Cambridge, Massachusetts 02139

ABSTRACT

Three types of analytical models for determining adequate feedback spatial and temporal resolution are distinguished for representing dynamics with finite sampling: piecewise continuous,
('spliced') discrete coupled modes, and coupled wavetrains. The
stabilization of the z-θ pinch is used for comparing the various representations, particularly as they represent a quasi-one-dimensional model for the m = 1 modes, and specific stability regimes are
given. The advantages of the modal and wavetrain approaches in
describing three-dimensional effects are illustrated, with the wavetrain approach shown as particularly convenient for systems having
many potentially unstable wavelengths within the system boundaries,
hence necessitating many sampling stations. A discussion of difficulties in stabilizing interchange modes is given, and nonlinear
forms of 'bang-bang' feedback obviating these difficulties suggested.

———————————————

I. CONTINUUM FEEDBACK SYSTEMS

From the viewpoint of processing information, crucial questions
for active control of plasmas are: how many feedback stations are
required, and where located? Because forcing electrodes must be
outside the plasma, what loss of resolution is associated with acting
at a distance on the plasma through quasi-static fields? From the
physical point of view, an equally important question is: what
'handles' are available for pushing the plasma about? A series of
examples is used here to answer some of these questions with the
m=1 modes of the z-θ pinch the major theme.

A.   The Pinch as a Feedback System

The dynamics of the z-θ pinch with feedback can be conveniently represented by the feedback network of Fig. 1.   The physical
configuration is sketched in Fig. 2, with the plasma modeled, for

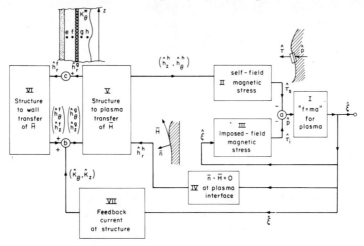

Fig. 1.   Schematic feedback representation of z-θ pinch shown
in Fig. 2.   The continuum transfer functions are summarized in
Table I.   Radial positions at which the amplitudes are evaluated are
indicated by superscripts and the inserted cross-sectional view of
the system at the upper left.

simplicity, as a perfectly-conducting incompressible fluid of mass
density ρ.   The field is excluded from the plasma and in equi-
librium is $H_\theta (R/r)\bar{i}_\theta + H_z \bar{i}_z$ in the vacuum.   Though the variables
can be variously regarded as Fourier transforms or mode ampli-
tudes, assume for now that they are merely the complex amplitudes
for solutions taking the form ξ = Re ξ exp $j(\omega t - m\theta - kz)$.

The radial deflection $\hat{\xi}$ is the 'system output', feeding three
feedback loops through III, IV, and VII in the diagram: these ulti-
mately determine the magnetic stress $\tau = \tau_s + \tau_i$ on the plasma-
vacuum interface, on node 'a'.   III represents the destabilizing
decrease in radial magnetic stress experienced by the interface as
it moves outward through the radially decreasing imposed H_θ field.
IV accounts for the perturbation in vacuum field $\bar{h}$ caused by the
conservation of flux through the interface.   In the absence of active
feedback, the vacuum fields are determined by this perturbation
field at the interface and flux conservation at the outer wall.   Here,

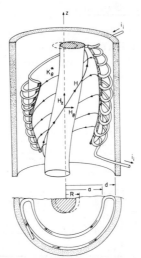

Fig. 2 Pinch with feedback structure showing part of one feedback station. A second coil also driven by K_θ^* but in the opposite
direction gives m = 1 dependence. Then a similar pair of coils is
interspersed with those shown, and rotated 90° to detect and drive
perpendicular components of deflection. Currents in the structure
are proportional to the plasma displacement opposite the center of
the feedback station. Station length in the direction of field lines
is ℓ^* and in the z- direction is ℓ. Length of plasma in z-direction
is L.

the vacuum region is broken into two parts, joined by field boundary
conditions at the feedback structure surface at r = a, as represented
by nodes 'b' and 'c'. Thus, the perturbation tangential field
($\hat{h}_z^h$, $\hat{h}_\theta^h$) at the interface is determined, and the "self-field" part of
the magnetic stress τ_s, at the interface, follows from II. The self-
consistent dynamics are represented by closing the feedback loop
and writing stress equilibrium at 'a' in terms of $\hat{\xi}$. The specific
transfer relations are summarized in Table I.

Active feedback is introduced by making the structure surface
current ($\hat{K}_\theta$, $\hat{K}_z$) proportional to the local deflection, thus driving
the jump in the tangential fields at the structure, represented by
'b' in the diagram. This current acts through V and VI, which act
as spatial filters, passing long wavelengths to augment h_z^h, h_θ^h and
hence contribute to the stress equilibrium at the interface. Combining the relations of Table I gives as a summation at 'a',
$\hat{p} = - \tau_i - \tau_s$, or,

Table I

Summary of transfer relations for z-θ pinch. Dimensions R, a, and d are defined in Fig. 2. Superscripts designate radial locations shown in upper left of Fig. 1. J_m and H_m are Bessel and Hankel functions.

I.	$\hat{\xi} = \dfrac{\hat{p}}{\omega^2 \rho R}\left[\dfrac{(jkR)J_m'(jkR)}{J_m(jkR)}\right]$
II.	$\hat{\tau}_s = -\mu_0[H_\theta(\frac{m}{kR}) + H_z]\,\hat{h}_z^h$
III.	$\hat{\tau}_i = \dfrac{\mu_0 H_\theta^2}{R}\,\hat{\xi}$
IV.	$\hat{h}_r^h = -jk[H_\theta(\frac{m}{kR}) + H_z]\,\hat{\xi}$
V*	$\begin{bmatrix}\hat{h}_z^g \\ -\hat{h}_z^h\end{bmatrix} = \dfrac{k}{m}\begin{bmatrix}a\hat{h}_\theta^g \\ -R\hat{h}_\theta^h\end{bmatrix} = \Delta^{-1}(a,R)\begin{bmatrix}F(a,R) & -F(a,a) \\ -F(R,R) & F(R,a)\end{bmatrix}\begin{bmatrix}\hat{h}_r^g \\ \hat{h}_r^h\end{bmatrix}$
VI*	$\hat{h}_z^f = \dfrac{ka}{m}\hat{h}_\theta^f = -\dfrac{F(a,d)}{\Delta(d,a)}\hat{h}_r^f$
VII	$\hat{K}_\theta = -\dfrac{ka}{m}\hat{K}_z = \hat{K}_\theta(\hat{\xi})$

* $F(\alpha,\beta) = H_m(jk\alpha)J_m'(jk\beta) - J_m(jk\alpha)H_m'(jk\beta)$

$\Delta(\alpha,\beta) = J_m'(jk\alpha)H_m'(jk\beta) - J_m'(jk\beta)H_m'(jk\alpha)$

$$\frac{\omega^2 \rho J_m(jkR)}{(jkR)J_m'(jkR)} \hat{\xi} = -\frac{\mu_o H_\theta^2}{R} \hat{\xi} + jk\mu_o \left[H_\theta \left(\frac{m}{kR} \right) + H_z \right]^2 .$$

$$\left[F(R,a) - \frac{F(R,R)}{F(a,R) + \frac{\Delta(a,R)F(a,d)}{\Delta(d,a)}} \frac{F(a,a)}{\Delta(a,R)} \right] \frac{\hat{\xi}}{\Delta(a,R)}$$

$$+ \quad \frac{\mu_o F(R,R)[H_\theta (\frac{m}{kR}) + H_z]\hat{K}_\theta}{F(a,R) + \frac{\Delta(a,R)F(a,d)}{\Delta(d,a)}} \tag{1}$$

B. Ideal Continuum of Feedback

In the limit in which there are infinitely many sampling stations per unit length, the feedback approaches a continuum and the feedback law, VII, simply makes the structure current a function of $\hat{\xi}$. For illustration, the law does not involve operations in space or time; structure current is proportional to the local plasma deflection and perpendicular to the equilibrium, $\overline{H}$.

$$K_\theta^* \equiv \frac{H_z}{H} K_\theta - \frac{H_\theta}{H} K_z = A\hat{\xi} \tag{2}$$

II. QUASI-ONE-DIMENSIONAL MODELS

A. z-θ pinch

In the long-wave limit (kR < ka < kd $\ll$ 1), Eq. (1) reduces to a polynomial in ($H_\theta m + kRH_z$) and ω, and we can infer a 'string' (or long-wave) model involving one space dimension z^* measured in the direction of the total imposed field (sketched in Fig. 3) and time. The quasi-one-dimensional equations of motion are summarized in Table II for the m = 1 mode. Field stiffening of perturbations propagating in the z^* direction accounts for the second derivative, hence plays the role of a string tension. The adverse curvature of H_θ gives a radial stress tending to increase with the displacement itself; the parameter N' is therefore a normalized destabilizing magnetic pressure. The normalized feedback gain is M'_, with the feedback current in the θ^* direction. For illustration K_θ^* is made uniform over a station length ℓ^* and proportional to the local plasma deflection opposite the station center, as sketched in Fig. 3.

Features of the model that recommend themselves are the natural stabilization of short wavelengths which propagate as Alfven

surface waves along H; N' tends to destabilize only the long wavelengths. Moreover, the feedback scheme does not strongly influence
the short wavelengths where finite sampling renders the feedback
signal as likely to produce instability as stability.

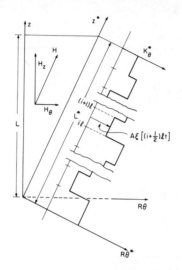

Fig. 3. Magnetic field coordinates and feedback scheme for
sampled system. For the quasi-one-dimensional model to be valid,
the sampling interval ℓ must exceed cross-sectional dimensions.

B. An Analog Case Study

The 'string' model takes a canonical form which has been the
basis for extensive studies of effects of sampling, equilibrium convection[*](Melcher, 1965a, 1965b, Crowley, 1967) etc. Figure 4
illustrates an experiment which is governed by completely analogous
equations of motion as for the long-wave m=1: equations summarized
in Table III. The string (or membrane) has tension T but is perfectly
conducting, hence unstable as V_o is raised to a threshold value. The
destabilizing normalized pressure, N', is analogous to that from H_θ.
In the experiment, feedback M' is used to project the system into a
regime of N', also sketched in Fig. 4, wherein it would otherwise be
unstable. Experimentally observed bounds on the regime of stability
come from limitations of the sampling; instability is exponential to
the right(the string slaps the electrodes) and oscillatory above (feedback regenerates short-wavelength oscillations).

---

[*]For a background on the 'electromechanical string', see J. R.
Melcher, IEEE Spectrum (October 1968), p. 86.

Table II

Summary of quasi-one-dimensional model for m = 1 modes of z-θ pinch.   Coordinates are defined in Fig. 3.

$$\frac{1}{v_A^2} \frac{\partial^2 \xi}{\partial t^2} = \frac{\partial^2 \xi}{\partial z*^2} + \frac{N'}{L*^2} \xi - \frac{M'}{L*^2} \left(\frac{K_\theta^*}{A} \right)$$

where

$$\xi = \text{Re } f(z,t)e^{-j\theta}$$

$$v_A^2 = \frac{\mu_o H^2}{\rho} \frac{[(\frac{R}{d})^2 + 1]}{[1 - (\frac{R}{d})^2]}$$

$$N' = \frac{H_\theta^2}{H_z^2} \frac{L^2}{R^2} \frac{[1 - (\frac{R}{d})^2]}{[1 + (\frac{R}{d})^2]}$$

$$M' = A \frac{L^2 H}{H_z^2 a} \frac{[1 - (\frac{a}{d})^2]}{[1 + (\frac{R}{d})^2]}$$

$$\begin{pmatrix} L* \\ \ell* \end{pmatrix} = \frac{H}{H_z} \begin{pmatrix} L \\ \ell \end{pmatrix}$$

$$H \frac{\partial \xi}{\partial z*} \equiv H_z \frac{\partial \xi}{\partial z} + \frac{H_\theta}{R} \frac{\partial \xi}{\partial \theta}$$

Previous work on the electromechanical string is directly applicable to the m = 1 modes of the pinch, at least in the long-wave limit. Three-dimensional effects are qualitatively indicated by the detailed results which have been obtained for the analogous case. Most important, the previous work indicates various points of view that can be taken in representing the feedback dynamics, and the following remarks pertain to either physical situation. In Sec. III we use variables for the z-θ pinch. Because three-dimensional effects are included in Secs. IV and V, the electric analog variables are used there.

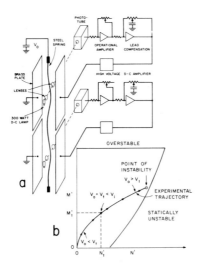

Fig. 4. (a) Analog feedback system. Equilibrium electric pressure due to V_O (N') causes instability stabilized by feedback M'. Equations of motion are summarized in long-wave limit by Table III. (b) Experimental trajectory for operation in regime that would be unstable in the absence of feedback.

III. PIECE-WISE REPRESENTATION OF SPATIAL SAMPLING

A representation of the sampling which is straightforward, but limited to quasi-one-dimensional situations, regards the motions as modes constructed by splicing solutions together at the boundaries between sampling stations. In the i'th sampling interval, we take $\xi = \text{Re } \hat{\xi}_i(z)e^{j\omega t}$ and the equations of motion become

$$[\frac{d^2}{dz*2} + k^2]\tilde{\xi}_i = \frac{M'}{L*2} \tilde{\xi}_i[(i+\frac{1}{2})\ell*]; \quad i\ell* < z* < (i+1)\ell*$$
$$(kL*)^2 = (\omega L*/v_A)^2 + N' \quad (3)$$

Solutions within the i'th station interval are therefore

Table III

Quasi-one-dimensional model for electromechanical string of Fig. 4; (d, ℓ, L) are, respectively, distances from string to electrodes between sampling points and between fixed ends; v is perturbation electrode potential.

$$\frac{1}{v_s^2} \frac{\partial^2 \xi}{\partial t^2} = \frac{\partial^2 \xi}{\partial z^2} + \frac{N'}{L^2} \xi - \frac{M'}{L^2} \left(\frac{v}{A}\right)$$

where

$$v_s^2 = \frac{T}{\rho_s} \quad ; \quad T = \text{tension}$$

$$\rho_s = \text{mass/length}$$

$$N' = \frac{2\epsilon_o V_o^2 L^2}{d^3 T}$$

$$M' = \frac{2\epsilon_o V_o A L^2}{d^2 T}$$

$$v = v(\xi)$$

$$\tilde{\xi}_i = C_i \cos kz^* + D_i \sin kz^* + G_i \qquad (4)$$

where the constants C_i, D_i and G_i in an s station system comprise 3s unknowns. The deflections are fixed at the ends, and since ξ and $d\xi/dz^*$ are continuous at the s-1 splicing points between sampling stations, we have 2s boundary conditions. There are s more conditions from the requirement that Eq. (4) satisfies (3). Hence, the problem reduces to a system of 3s homogeneous transcendental equations which define the eigenfrequencies ω. Detailed work has been reported (Melcher 1965a) for one-, two-, three- and four-station systems, with stability regimes found as illustrated in Fig. 5.

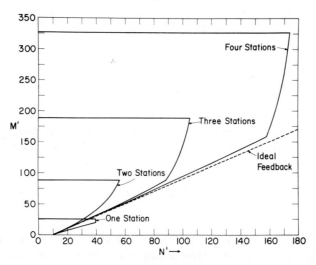

Fig. 5. Stability regime in the feedback-H_θ magnetic pressure plane for z-θ pinch and analog showing effect, predicted by piecewise continuous representation, of increasing number of stations with total length fixed. While static instability results to the right, oscillations result above the enclosed regimes.

Overstability is the result of too much feedback gain M' while static instability results from too much H_θ magnetic pressure N'. As expected, ideal feedback conditions are approached as the number of stations is increased, with the abrupt limit to the right caused by a sampling interval that approaches a critical half-wavelength.

IV. MODAL REPRESENTATION OF SPATIAL SAMPLING

The most popular representation (Melcher, 1965a; Gould, 1969) is in terms of modes for the system without feedback; for our example, simply Fourier modes

$$\xi = \text{Re } \tilde{\xi}(z)e^{j\omega t} \quad ; \quad \tilde{\xi} = \text{Re} \sum_{m=1}^{\infty} \Xi_m \sin k_m z; \quad k_m = \frac{m\pi}{L} \tag{5}$$

The feedback at the structure is represented in a roundabout fashion in terms of the same modes. In the case of Fig. 4, the electrode potentials (analogous to the surface currents K_θ^*) are

$$v = \sum_{n=1}^{\infty} V_n \sin k_n z \tag{6}$$

where

$$V_n = \frac{2A}{\pi n} \sum_{i=1}^{s} \tilde{\xi}[(i-\tfrac{1}{2})\ell] [\cos \frac{(i-1)n\pi}{s} - \cos \frac{in\pi}{s}]$$

$$\tilde{\xi}[(i-\tfrac{1}{2})\ell] = \sum_{q=1}^{\infty} \Xi_q \sin \frac{q\pi}{s}(i-\tfrac{1}{2})$$

Substitution of Eqs. (5) and (6) into the equations of motion shows that the feedback couples the modes. The equation of motion is satisfied for all z only if the coefficients of each $\sin k_m z$ vanish and these take the form:

$$\sum_{n=1}^{\infty} \Xi_n [(\Omega_m^2 - \Omega^2)\delta_{mn} + M'F_{mn}] = 0 \tag{7}$$

where $\Omega_m = \Omega_m(N')$ and F_{mn} accounts for the particular sampling and forcing scheme. Truncation of the number of modes used to represent the system with feedback at n = q is consistent with writing Eq. (7) for m = 1···q. Hence, a homogeneous system of equations in the mode amplitudes Ξ_n is obtained. The compatibility condition determines the normalized eigenfrequencies Ω in terms of the normalized uncoupled eigenfrequencies $\Omega_m(N')$ and the normalized feedback gain, M'.

Typical results (Melcher, 1965a) using modes are sketched in Fig. 6 for a four-station system. An advantage of the modal representation is that it represents three-dimensional effects. Thus, included in Fig. 6 is the blurring effect of the finite distance d between structure and string. The d/L = 0 case should be compared to the four-station result of Fig. 5. In general, 3s modes are required to describe reasonably the sampling limitations on the stability regimes, so the order of the eigenvalue equation is the same in the two cases. However, the modal approach requires solution of a polynomial eigenfrequency equation, while the splicing approach

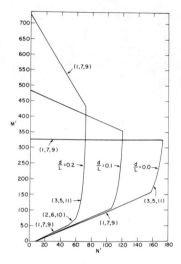

Fig. 6. Stability regimes predicted by coupled modes for four-station system. Blurring effect of finite electrode-string spacing d is illustrated for the analog case. Critical coupled modes are indicated by sets of three numbers.

gives a transcendental expression.

B.   Stabilization of a Rayleigh-Taylor Instability

A typical physical situation (Melcher and Warren, 1966) in which the modal approach is appropriate is shown in Fig. 7, wherein an interface of water is made unstable by applying an electric stress and the equilibrium is stabilized by feeding back to adjacent electrodes which both detect the motion and force the electric stress. The modes are combinations of Bessel functions that satisfy a fourth-order equation and four boundary conditions. Also, the detection is of the average deflection, which has the advantage of avoiding feedback-induced overstabilities evident in Figs. 5 and 6 because of the point-detection scheme.

V.   WAVETRAIN REPRESENTATION OF SAMPLING IN INFINITE SYSTEMS

With ideal continuum feedback, the stability problem amounts to simply inserting the feedback law for $\hat{K}_\theta$ into Eq. (1) and treating the coefficient of ξ as a dispersion equation. A powerful method of dealing with spatially sampled systems is the generalization of the dispersion equation viewpoint (Melcher, 1965a). Using the one-dimensional examples for discussion, we have

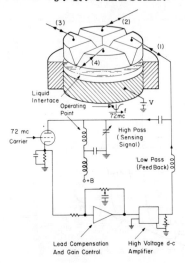

Fig. 7. Configuration for detailed experimental and theoretical studies using modal representation. Feedback stabilizes the Rayleigh-Taylor instability induced by the electric field.

$$
\begin{bmatrix} \Xi(k) \\ V(k) \end{bmatrix} = \int_{-\infty}^{\infty} \begin{bmatrix} \tilde{\xi}(z) \\ \tilde{v}(z) \end{bmatrix} e^{-jkz} \, dz \tag{8}
$$

and Fourier techniques show that the feedback law with sampling is

$$
V(k) = \frac{2A}{k\ell} \sum_{i=-\infty}^{\infty} \Xi(k - \frac{2\pi i}{\ell}) \sin\left(\frac{k\ell}{2} - \pi i\right) \tag{9}
$$

Thus, instead of the dispersion equation, from the equation of motion we obtain a coupling relation between wavetrains.

$$
\Xi(k) \{ [\Omega(N,k)]^2 - \Omega^2 \} + M \sum_{i=-\infty}^{\infty} F_i(k) \, \Xi\left(k - \frac{2\pi i}{\ell}\right) = 0 \tag{10}
$$

To examine the stability of a wavetrain, consider one characterized by a wavenumber β_0 but involving 'harmonics', so that Eq. (10) can be satisfied.

$$
\Xi(k) = \sum_{m=-\infty}^{\infty} \{ A_m \delta[k - (\beta_0 + \frac{2\pi m}{\ell})] + A_m^* \delta[k + (\beta_0 + \frac{2\pi m}{\ell})] \} \tag{11}
$$

With substitution of (11) into (10), coefficients of like argument delta functions, δ, must vanish; for argument $k - (\beta_0 + 2\pi q/\ell)$,

$$A_q'\{[\Omega_q(N,\beta_0,q)]^2 - \Omega^2\} + M \sum_{m=-\infty}^{\infty} A_m G_{mq}(\beta_0 \ell) = 0 \qquad (12)$$

Thus, an expression is obtained for the coupling of wavetrain amplitudes. Truncation with a finite number of amplitudes gives a system of equations homogeneous in the amplitudes A_m, expressions that allow the calculation of the eigenfrequency associated with a given real β_0. All real values of β_0 are examined for stability in a manner familiar from dispersion-equation theory. Stability regimes are summarized in Fig. 8, using a system of three amplitudes.

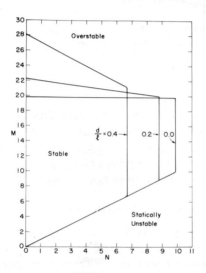

Fig. 8. Stability regimes predicted by traveling-wave representation; $d/\ell = 0$ pertains to z-θ or analog case, $d/\ell \neq 0$ for analog only. $N = N'(\ell/L)^2$, $M = M'(\ell/L)^2$.

The wavetrain approach gives stability regimes in the limit where the overall length $L \rightarrow \infty$, while the sampling interval ℓ remains finite. It has the advantage of representing three-dimensional effects, even if the uncoupled eigenmodes are unknown. For example, its application to the general system of Fig. 1 is straightforward, with the amplitudes then representing Fourier transforms and the feedback law, like Eq. (9), including a sampler. Detailed work on hydromagnetic instability following this approach has been reported for the Rayleigh-Taylor instability, as sketched in Fig. 9 (Melcher, 1966). Recent developments (Dressler 1970) show how a closed-form solution of the sampled system is possible, and how such a viewpoint is useful in representing temporal sampling and combined time-space sampling (scanning).

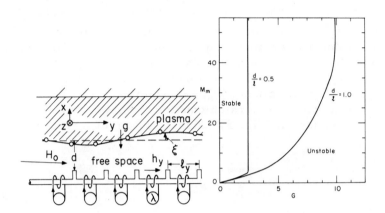

Fig. 9. Study of Rayleigh-Taylor instability following wave-train approach. System is stable in regime above lines shown; $G = d^2 g \rho / T$ and $M_m = d^2 \mu_o H_o A / T$, where T is the surface tension. Feedback field, $h_y = A < \xi >$, is proportional to the average of the deflection over the station area, hence no overstability is found.

VI. INTERCHANGE MODES: A DILEMMA FOR LINEAR FEEDBACK

From Eq. (1) it is clear that, for interchange modes wherein $H_\theta / H_z = -kR/m$, the feedback can have no stabilizing effect. There are two underlying physical reasons. First, the only perturbation $\bar{H}$ that can produce a linear stress at the interface is in the direction of imposed $\bar{H}$. Second, magnetic perturbations of the vacuum field are irrotational, so that $(\hat{h}_z / \hat{h}_\theta) = kR/m$. Thus,

$$H_\theta / H_z = -kR/m = -\hat{h}_z / \hat{h}_\theta , \tag{13}$$

hence the perturbation field must be perpendicular to $\bar{H}$. Regardless of the feedback law, linear feedback has no influence on the interchange mode.

With finite boundaries, the interchange modes can be avoided by excluding them from modes allowed by boundary conditions. For example, with the ends fixed as discussed here, two modes are required to satisfy boundary conditions: if one is in the interchange direction, the other is not.

It is easy to cover up the catastrophe that occurs in the exchange

direction.   Feedback to K_θ^* (Table II) causes no apparent dilemma
in the interchange direction.   But, if K_θ^* is finite and perturbations
are purely in the interchange direction, then K_θ must be infinite,
a requirement inconsistent with the linearized model.   Similar
observations have been made before in the literature (Melcher, 1966,
Canales, 1965).

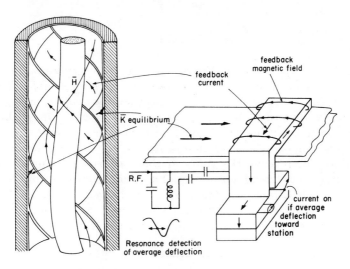

Fig. 10.   Non-linear feedback scheme.   Feedback currents are
parallel to the imposed field; on if plasma motions are toward
station, and off if they are away.   Detection of average motions
might be achieved be using the driving structure with deflections
measured by the modulation of a high-frequency superimposed
signal.

Nonlinear feedback to fields perpendicular to the imposed field
offer a solution to the problem.   Figure 10 sketches a possible configuration for the z-θ pinch.   Feedback currents are either 'on',
adjacent to plasma interface sections that approach the wall, or
'off'.   This type of 'bang-bang' stabilization is currently under investigation (Millner and Parker, 1970).

ACKNOWLEDGMENT

Motivation for our observations on the z-θ pinch stems in part
from discussions with F. L. Ribe and G. A. Sawyer of Los Alamos
Scientific Laboratory Project Sherwood Group, and from suggestions
from R. R. Parker and K. I. Thomassen of M.I.T.   This work was
supported by NASA Grant NGL-22-009-014.

2.2 NONLINEAR STABILIZATION OF A CONTINUUM

A. R. Millner and R. R. Parker
Department of Electrical Engineering and Research Laboratory
of Electronics, Massachusetts Institute of Technology,
Cambridge, Massachusetts 02139

ABSTRACT

Nonlinear, or "bang-bang," feedback, in which a constant corrective force of arbitrary strength is applied to the plasma for as long a time as the local average of the surface displacement is positive, is discussed for the stabilization of MHD modes. It is shown that, in addition to obvious bandwidth and impedance-level problems, linear feedback cannot stabilize modes for which the perturbation amplitude is constant along the lines of the external magnetic field. Plasma stability with "bang-bang" feedback scheme is studied using an energy principle (Liapunov function). Although we focus attention on modes whose amplitude is constant along B_o, the proposed scheme can be applied to stabilize all other modes as well.

I. INTRODUCTION

The use of feedback in order to stabilize MHD modes in Tokamak configurations is of great interest, since its success would allow an increase in the induced current and an attendant increase in the heating rate and confinement. In this paper, we first examine the possibility of using linear feedback and show that, in addition to practical problems, linear feedback cannot stabilize modes whose amplitudes are constant along the lines of the equilibrium magnetic field. We then explore the possibility of employing nonlinear or "bang-bang" feedback, and show that, in addition to being somewhat easier to implement, this method has no inherent shortcomings regarding these modes.

II. LINEAR FEEDBACK

We consider first the problem of applying linear feedback to a long, straight plasma column of radius a, length L, and carrying a current I on its surface. The ideal-MHD equations yield the following equation for the linearized displacement, $\vec{\xi}$:

$$\frac{\partial^2 \vec{\xi}}{\partial t^2} - c^2 \nabla\nabla \cdot \vec{\xi} + \frac{c_H^2}{B_i^2} [\vec{B_i} \times \nabla\times(\nabla\times\vec{\xi}\times\vec{B_i})] = 0 \quad , \tag{1}$$

where c is the sound speed, c_H the Alfven speed, and $\vec{B_i}$ is the equilibrium field inside the column (parallel to its axis).[1] Normal modes, proportional to $\vec{\xi}_\kappa(r)e^{i\omega_\kappa t}$ are thus determined by the equation

$$\omega_\kappa^2 \vec{\xi}_\kappa + c^2 \nabla\nabla \cdot \vec{\xi}_\kappa - \frac{c_H^2}{B_i^2} [\vec{B_i} \times \nabla\times(\nabla\times\vec{\xi}_\kappa\times\vec{B_i})] = 0 \tag{2}$$

together with boundary conditions at the plasma-vacuum interface,

$$[\vec{B} \cdot \hat{n}]_i = [\vec{B}\cdot\hat{n}]_o \tag{3}$$

$$[\hat{n} \cdot \overline{\overline{T}} \cdot \hat{n}]_i = [\hat{n} \cdot \overline{\overline{T}} \cdot \hat{n}]_o \quad , \tag{4}$$

where $\hat{n}$ is the normal to the perturbed surface, and $\overline{\overline{T}}$ is the stress tensor, $\overline{\overline{T}} = (1/\mu_o)\vec{B}\,\vec{B} - (1/2\mu_o) B^2\overline{\overline{\delta}} - p\overline{\overline{\delta}}$. The brackets denote the operation of taking first-order quantities and the subscripts i and o signify points just inside and outside the perturbed surface. The eigenvectors and dispersion relation generated by Eqs. (2) - (4) have been given by Shafranov (1957).

We are interested in feedback control of the unstable modes. Consequently, we assume that specified currents exist in the vacuum region which will subsequently be controlled in order to effect stabilization. These currents give rise to an external field, $\vec{H}_{ex}(\vec{r},t)$, which we shall define as having zero normal component at the unperturbed plasma surface.

The plasma response can be found by expanding in the eigenvectors of the normal modes. Hence, we write

$$\vec{\xi} = \sum_\kappa a_\kappa(t)\, \xi_\kappa(\vec{r}) \quad .$$

Since ω_κ^2 is real, $\vec{\xi}_\kappa$ can also be chosen real. Thus, if $\vec{\xi}'_\kappa$ is a complex eigenvector proportional to $\exp[i(2n\pi/L) z - im\theta]$, we can construct two linearly independent and orthogonal eigenvectors by taking the real and imaginary parts of $\vec{\xi}'_\kappa$. By manipulating Eq. (2) written for κ and κ' the eigenvectors can be shown to be orthogonal, $\int \vec{\xi}_\kappa \cdot \vec{\xi}_{\kappa'} \, d\tau = \delta_{\kappa\kappa'} \int \xi_\kappa^2 \, d\tau$, and therefore

$$a_\kappa = \frac{\int \vec{\xi} \cdot \vec{\xi}_\kappa \, d\tau}{\int \xi_\kappa^2 \, d\tau} \quad .$$

The equations of motion for the a's can be found by dotting Eq. (2) with $\vec{\xi}_\kappa$, Eq. (1) with $\vec{\xi}$, adding and integrating over the unperturbed plasma volume. After two integrations by parts and use of the boundary conditions, we get

$$\frac{d^2 a_\kappa}{dt^2} + \omega_\kappa^2 a_\kappa = -\frac{\int \vec{H}_{ex} \cdot \vec{B}_o \, \vec{\xi}_\kappa \cdot d\vec{S}}{\int \rho_o \xi_\kappa^2 \, d\tau} \quad . \tag{5}$$

The surface integration on the right-hand side is over the unperturbed plasma surface, and $\vec{B}_o = B_z \hat{i}_z + B_\theta \hat{i}_\theta$ is the equilibrium field just outside the surface.

A dilemma of linear feedback can now be appreciated by supposing that $q = aB_z/RB_\theta$, where $R = L/2\pi$, is rational at the surface. Then, for a mode with $m/n = q$, i.e., one which is constant along $\vec{B}_o$, the right-hand side of Eq. (5) must vanish identically. To see this, suppose, for example, that $q = 1$, and imagine carrying out the surface integration along strips which start at $z = 0$ and which are parallel to $\vec{B}_o$. Since $\vec{\xi}_\kappa$ and $\vec{B}_o$ are constant on each strip, the contribution of a given strip is proportional to $\int \vec{H}_{ex} \cdot d\vec{l}$. For a system with strict periodicity, as in a torus, this line integral must be independent of the initial value of θ. Hence the θ-integration involves only $\vec{\xi}_\kappa(\theta, z = 0) \sim \cos m\theta$ and must vanish.

An additional problem with linear feedback is the practical problem of controlling the required field over a bandwidth substantially larger than the growth rate (typically 1 - 10 MHz). As Wang (1970) has shown, complete stability could be obtained for $B_{ex} = KB_o (\xi_r/a)$ providing $K > B_\theta^2/B_o^2 = (a/qR)^2$. (Note, however, that this field is necessarily nonphysical in view of our previous argument.) Control of a perturbation equal to 10% of **a** would require a field, even in this idealized case, of the order of $10^{-3} B_o$ which for Tokamaks may be 100 G. Control of fields this large over a 10 - 100 MHz bandwidth is obviously a discouraging prospect.

III. NONLINEAR FEEDBACK

We now examine the prospects of nonlinear or "bang-bang" feedback. The strategy here is to apply a <u>constant</u> corrective force or arbitrary strength to the plasma for as long a time as the local average of the surface displacement is positive. For example,

Fig. 1 shows a perfectly-conducting fluid supported against gravity

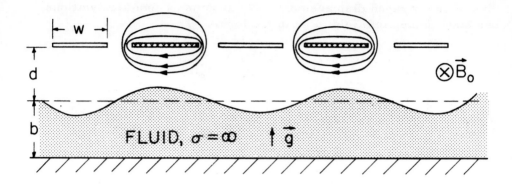

Fig. 1. Example of bang-bang feedback applied to a Rayleigh-Taylor unstable fluid. The fluid is perfectly conducting, incompressible, and inviscid, and is supported against gravity by a self-magnetic field. Large-amplitude stabilizing currents are applied to those straps under which the average displacement is positive.

by a self-magnetic field. The system is Rayleigh-Taylor unstable. Large-amplitude stabilizing currents would then be switched on in those straps under which the fluid has positive displacement. When the displacement crosses to negative values, the current is switched to zero. We remark that, from a practical viewpoint, the switching of sizable currents in times which are short compared to growth rates seems to be more reasonable than the analogous requirement for linear feedback.

The preceding formalism can be applied to this problem without difficulty and the equation for the normal coordinate, a_κ, becomes

$$\ddot{a}_\kappa + \omega_\kappa^2 a_\kappa = -\frac{1}{2\mu_o}\frac{\int B_{ex}^2 \vec{\xi}_\kappa \cdot d\vec{S}}{\int \rho_o \xi_\kappa^2 \, d\tau}. \tag{6}$$

In obtaining Eq. (6), we have kept only terms which are linear in $\vec{\xi}$ and have, for simplicity, focussed attention on those modes for which $m/n = q$. For $m/n \neq q$, the right-hand side of Eq. (5) must be added to the right-hand side of Eq. (6). In addition, we have neglected curvature of B_{ex}^2 in comparison with that of B_o^2.

The applied field is switched on and off in accordance with the sign of an averaged displacement. We assume N feedback stations and take the applied field to be in the form

$$B_{ex}^2 = \sum_{\ell}^{N} \frac{1}{2} \left(1 + \frac{D_\ell(t)}{|D_\ell(t)|}\right) B_\ell^2(\vec{r}) \tag{7}$$

where D_ℓ, the "discriminant" associated with the ℓth station, is given by

$$D_\ell(t) = \frac{\int B_\ell^2(\vec{r}) \vec{\xi} \cdot ds}{\int B_\ell^2(\vec{r}) \, dS} . \tag{8}$$

Except for a translation of coordinates, $B_\ell^2(\vec{r})$ is assumed to be the same for all stations.

To fix ideas, consider again the system in Fig. 1. For $d \ll w$, B_ℓ^2 can be approximated as being a constant, B_o^2, under the ℓth strap and zero otherwise. Let the normal component of the displacement at the boundary be $\Xi(z,t) = \sum_n \Xi_n' \cos k_n z + \sum_n \Xi_n \sin k_n z$. The equation for Ξ_n is easily found to be

$$\frac{\rho_o}{k_n \tanh k_n b} \ddot{\Xi}_n + \lambda_n^2 \Xi_n = -\frac{B_o^2}{2\mu_o} \sum_{\ell=1}^{N} A_{\ell n} \left(1 + \frac{D_\ell}{|D_\ell|}\right) \tag{9}$$

where $k_n = 2n\pi/L$, $\lambda_n^2 = -\rho g + T k_n^2$, T being a surface tension, $D_\ell = \int_{R_\ell} \Xi(z,t) \, dz$ and $A_{\ell n} = \int_{R_\ell} \cos k_n z \, dz$. The equation for Ξ_n' is the same, with $\sin k_n z$ replacing $\cos k_n z$ in the definition of $A_{\ell n}$. With proper identification, Eq. (9) is seen to have the same form as Eq. (6).

A stability analysis for either Eq. (9) or (6) can be formulated in terms of an energy principle (Liapunov function). Multiplying Eq. (6) by $\dot{a}_\kappa$ and also by X_κ, the denominator on the right-hand side, and summing over all modes, yield $dE/dt = 0$, where $E = T + \psi + U$ with

$$T = \frac{1}{2} \sum_\kappa X_\kappa \dot{a}_\kappa^2 \geq 0, \quad \psi = \sum_\kappa X_\kappa \omega_\kappa^2 a_\kappa^2 ,$$

and

$$U = \frac{1}{2} \sum_\ell (D_\ell + |D_\ell|) \left(\int \frac{B_\ell^2}{2\mu_o} \, dS \right)$$

$$\simeq \frac{1}{2} \left(\int \frac{B_\ell^2}{2\mu_o} \, dS \right) \sum_\ell |D_\ell| \quad .$$

In the last approximation, we assume that the feedback stations are densely packed, so that if all currents are "on," B_{ex}^2 will be sensibly constant, and will have negligible coupling to any mode, except possibly m = 0, n = 0.

Let $\vec{S} = (\dot{a}_1, \dot{a}_2, \ldots; a_1, a_2 \ldots)$. Then a sufficient condition for stability is $\nabla E \cdot \vec{S} > 0$ for arbitrarily small $|\vec{S}|$. This condition is clearly satisfied for all directions in state-space except those for which U = 0, because T and U are positive definite and $U > |\psi|$ for sufficiently small $|\vec{S}|$. Since the condition U = 0 imposes N independent constraints on the components of $\vec{S}$, a system of N feedback stations is seen necessary in order to stabilize N unstable modes.

ACKNOWLEDGMENT

The authors wish to thank Professor J. R. Melcher for providing impetus to this work and for many stimulating discussions. This work was supported in part by the U. S. Atomic Energy Commission [Contract AT(30-1)3980].

2.3 VIDEOTYPE SAMPLING IN THE FEEDBACK STABILIZATION OF ELECTROMECHANICAL EQUILIBRIA

John L. Dressler
Department of Electrical Engineering
Massachusetts Institute of Technology
Cambridge, Massachusetts 02139

ABSTRACT

The feedback control of hydromagnetically contained plasmas with dimensions large compared to potentially unstable wavelengths requires a large number of spatially distributed feedback sensors and drivers. The multiplicity of signals to be amplified and processed suggests the use of computers or other discrete-time devices which handle signals on a 'time-sharing' basis. Typically, scanning techniques are envisioned to sense and drive, thus introducing to an analytical representation discreteness in both time and space. A general method, based on the Fourier superposition of wavetrains, is developed to describe infinite continuum systems with discrete spatial and temporal feedback. Dynamics are represented by a generalization of the dispersion equation, with Z transforms used to provide closed-form expressions if the discreteness is in space or in time only. The Bers-Briggs criterion is generalized to differentiate between absolute instabilities and amplifying waves with the discrete feedback. A quasi-one-dimensional model for the m = 1 mode of the z-θ pinch is studied to delineate effects of spatial and temporal sampling rates on stability regimes.

I. INTRODUCTION

For continuum feedback control systems which contain many feedback drivers and sensors, a time-sharing technique is attractive to generate the numerous control signals. At present, continuum feedback controls are spatially discrete systems; time-sharing will make them also temporally discrete systems. A previous analysis of spatially and temporally sampled control systems (Thomas, 1967) used a normal mode analysis which was too complicated to be useful when more then a few feedback stations were considered. A Fourier integral analysis, which ignores boundary conditions, provides an accurate description of spatially discrete systems with a large number of feedback stations and is much simpler than a normal mode analysis (Melcher, 1965a). This technique will be extended to

provide a closed-form representation and include the effects of
temporal sampling.

The system which has been studied is shown schematically in
Fig. 1. A continuum, described by

$$\frac{1}{v_p^2} \left(\frac{\partial}{\partial t} + U \frac{\partial}{\partial x} \right)^2 \xi(x,t) = \frac{\partial^2 \xi}{\partial x^2}(x,t) - \delta \frac{\partial \xi(x,t)}{\partial t} + N\xi(x,t) + F(x,t)$$

(1)

(which could be a plasma column in a pinch experiment, a jet of
water, or a string) is driven by a set of feedback stations, represented by the flat 'electrodes' in Fig. 1. The feedback stations

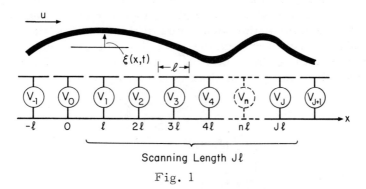

Scanning Length Jℓ

Fig. 1

exert a constant force to the length ℓ of the continuum adjacent to
the electrode. Each station is driven by a control signal, V_n, which
is set by the control system. These individual signals are adjusted
on a discrete-time basis with a period of T seconds. These feedback stations are also grouped into scanning lengths of J stations.
The control signals in these scanning sections are adjusted in a se-
quential manner from left to right. With this type of discrete temporal control, the control signal which drives the continuum for
$n\ell - \ell/2 < x < n\ell + \ell/2$ is then adjusted at the times $t = t_o + nt_o/J$,
$2t_o + nt_o/J$, $3t_o + nt_o/J$ . . . . . . The motivation for this control
scheme is the common television system which transmits information about a spatially distributed variable on a discrete-time basis.

Any real system that could be built would probably consist of only one scanning length. By placing an infinite number of these physical systems end to end, a model has been generated whose boundaries are at infinity, and a Fourier-LaPlace transform, shown in Eq. (2), can be used to describe the deflections of the system.

$$\xi(x,t) = \int_{-\infty}^{\infty} \int_{-j\infty}^{\infty} \frac{\Xi(S,K)}{(2\pi)^2 j} e^{St} e^{jKx} dK\, dS \tag{2}$$

II. DISPERSION EQUATION FOR SPATIALLY DISCRETE SYSTEMS

The system is first considered to be a continuous time control system, that is, each control signal V_n will be adjusted continuously in time. The spatially discrete driving force $F(x,t)$ generated by the feedback stations is shown in Fig. 2, as a serie of steps with discontinuities located at the boundaries of the feedback stations. This spatially discrete force can be considered as the result of a train of impulses along the x axis, $\phi^*(x,t)$, being fed into a spatial sample-and-hold filter. This filter is the first box in Fig. 2.

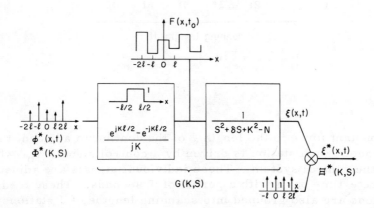

Fig. 2

The force generated by the drivers is applied to the continuum, which is represented for the case of no convection (U = 0) by the normalized transform of the Green's function in the second box, Fig. 2. The output of this second box is the displacement $\xi(x,t)$.

The control system measures ξ only at the center of each feedback station, but could just as well measure an average. A representation of the measured value of the displacement can be obtained by multiplying the displacement by a spatial impulse train. The resultant impulse train, $\xi*(x,t)$, is zero except at the points where $\xi(x,t)$ is measured. The impulse area of $\xi*(x,t)$ is the value of the displacement at the measuring or 'sampling' point.

From Z-Transform theory (Jury 1964) it is known that a transfer function $G*(K,S)$ exists which relates the input of Fig. 2 to the output:

$$\Xi*(K,S) = G*(K,S) \ \Phi*(K,S) \tag{3}$$

The formula for this discrete transfer is

$$G*(K,S) = G(K,S) \otimes \frac{1}{1-e^{-jK\ell}} \tag{4}$$

where $\otimes$ refers to convolution in K space. Evaluating Eq. (4) gives

$$G*(D,S) = \frac{e^{-\frac{1}{2}(S^2+\delta S-N)^{\frac{1}{2}}}}{2(S^2- N + \delta S)} \cdot$$

$$\left\{ \frac{(D-1)^2[1+e^{-(S^2+\delta S-N)^{\frac{1}{2}}}]}{[D(1+e^{-2(S^2+\delta S-N)^{\frac{1}{2}}}- (D+1)e^{-(S^2+\delta S-N)^{\frac{1}{2}}}]} \right\} + \frac{1}{S^2+\delta S - N} \tag{5}$$

where $D = e^{-jK\ell}$ is the discrete spatial wavenumber.

A method of stabilizing the system is to multiply the displacement by a negative gain and feed this signal to the input. This process is represented by the servo-loop in Fig. 3. The transfer function for the closed-loop system is given by Eq. (6).

$$H*(D,S) = \frac{G*(D,S)}{1 + AG*(D,S)} \tag{6}$$

Setting the denominator of Eq. (6) to zero gives the discrete dispersion equation from which the natural frequencies of the system are found in terms of the discrete wavenumber, **D**.

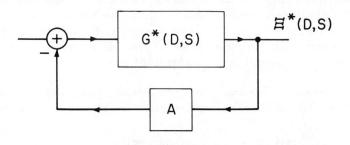

Fig. 3

III. STABILITY CRITERION FOR SPATIALLY DISCRETE SYSTEMS

A method of distinguishing between amplifying and evanescent waves, and for detecting the presence of absolute instabilities, has been developed by Bers and Briggs, (Briggs 1964) for systems which ar uniform in one spatial dimension, and the relation between the frequency S and the wavenumber K is given by the dispersion equation

$$\Delta(S,K) = 0 \tag{7}$$

This criterion cannot be used when there is discrete spatial feedback, because a simple dispersion equation like Eq. (7) does not exist. From Eq. (6), a dispersion relation between the frequency S and the discrete spatial wavenumber D can be found in the form

$$\Delta(S,D) = 0 \tag{8}$$

The Bers-Briggs criterion can now be used to distinguish the type of instability.

The inversion formula for finding the displacement at one of the sampling stations is

$$\xi(n\ell,t) = \frac{1}{2\pi j} \int_{-j\infty}^{j\infty} [\, \frac{1}{2\pi j} \int_{C} \Xi^*(D,S)D^{-(n+1)}dD \,]\, e^{St}dS \tag{9}$$

The system is driven by a source that is localized near the origin
of the x axis.   The poles of $\Xi^*(D, S)$ in the D plane will therefore be
the same as those of the closed-loop transfer function; that is, they
can be found from Eq. (8).   If the growth rate, σ , of the driving
source frequency, shown in Fig. 4a as an x, is sufficiently large,
then all waves will decay in space away from the origin.

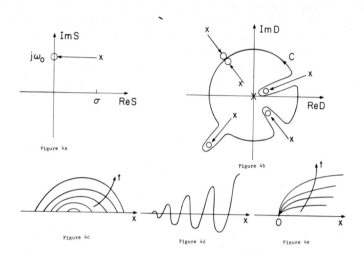

Fig. 4

The correct contour C to use on the inner integral of Eq. (10) for
this frequency is the unit circle.   The poles outside the unit circle
correspond to waves which decay to the right and those inside the
unit circle, excluding the pole of order n + 1 at the origin, correspond
to waves decaying to the left.

When the temporal growth rate of the source is decreased to
zero, the poles in the D plane (Fig. 4b) will move.   If two poles,
such as those in the second quadrant of Fig. 4b, pinch the contour
C between them, an absolute instability (shown in Fig. 4c) will
occur.   At every point, the displacement grows in time.   In the third
and fourth quadrants of Fig. 4b, the poles move across the unit
circle and deform contour C.   This causes a convective instability,
shown in Fig. 4d.   For positive values of n, the contour can be
pinched between the stationary pole at the origin and a moving pole.
This leads to an 'absolute convective' instability, shown in Fig. 4e.
This absolute instability, which does not propagate upstream, is a
consequence of spatial discreteness and has been observed by
Crowley (1967).

IV. DISPERSION EQUATION FOR A SPATIALLY AND TEMPORALLY DISCRETE SYSTEM

When the feedback sensors are scanned and the feedback signals adjusted in a discrete time manner, a force $F(x, t)$ illustrated in Fig. 5 is generated. The discontinuities on the x axis correspond to the edges of the feedback forcers; the discontinuities on the t axis correspond to the discrete times that the feedback signal is changed. This driving force can be considered as the result of an impulse array $\Phi^*(x, t)$ being fed into a space and time sample-and-hold filter. The impulses are separated by a time T on the t axis and a length ℓ on the x axis. There is a delay on the t axis corresponding to the position of a driver in a scanning length. The filter is the first box in Fig. 5.

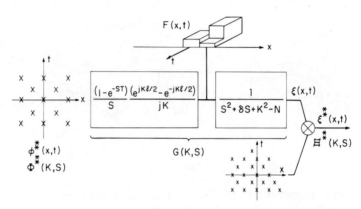

Fig. 5

The force is then applied to the continuum system, which is represented for the case of no convection, by the normalized transform of the Green's function (second box, Fig. 5). The control system will measure the continuous displacement, $\xi(x, t)$ only at certain points in the x-t plane. This process can be thought of as multiplying ξ by an impulse array. The resulting impulse array $\xi^*(x, t)$ represents the measured values of the displacement in the x-t plane.

The relation $G^*(K, S)$ between the input and output of Fig. 5 is

$$G^*(K,S) = G(K,S) \otimes \frac{1}{(1-e^{-ST})(1-e^{-ST/J} e^{-jK\ell})} \qquad (10)$$

where $\otimes$ refers to convolution in S and K space. Evaluating (10) gives

$$G_*^*(D,S) = \frac{1-e^{-ST}}{T} \sum_{n=-\infty}^{\infty} \frac{G*(De^{\frac{2n\pi}{J}}, S + \frac{2n\pi}{T})}{(S + \frac{2n\pi}{T})} \tag{11}$$

If the measured output of this system is multiplied by a negative gain, and returned to the input, a servo-loop diagram like Fig. 3 can be used to describe the system and a closed-loop transfer function like Eq. (6) results. The equation for the roots of this transfer function

$$1 + AG_*^*(D,S) = 0 \tag{12}$$

is the discrete dispersion equation for the system.

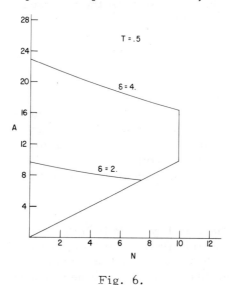

Fig. 6.

The dispersion relation Eq. (12) was examined for stability using the discrete stability criterion of Sec. III; the results are shown in Fig. 6. It was found that the series of Eq. (12) could be truncated with n = -1,0,1. The region of stability is bounded from above by an overstability caused by the discrete temporal and spatial nature of the feedback, on the right by the inability of the discrete spatial sampling to detect certain spatial modes, and on the bottom by the system's inherent static instability.

Acknowledgment

This work was supported by NASA Grant NGL-22-009-014.

2.4 FEEDBACK CONTROL OF KRUSKAL-SHAFRANOV MODES[*]

J. F. Clarke and R. A. Dory
Oak Ridge National Laboratory
Oak Ridge, Tennessee 37831

ABSTRACT

Kruskal-Shafranov modes on a cylindrical plasma are examined with respect to possible active feedback stabilization. Active stabilization of these modes, which would make tokamak reactors more realizable, is found to be possible in principle but to present great practical difficulties. The application of the technique to a reactor-type plasma will be considered.

I. INTRODUCTION

Present tokamak experiments are possibly limited in current by the appearance of MHD modes of the Kruskal-Shafranov type. We have analyzed these modes in a cylindrical geometry to find their susceptibility to feedback control by externally applied magnetic fields. We find that, within the limits of our model, this control is possible in principle but presents great practical difficulties.

II. THEORY

We have analyzed the stability of an infinite cylindrical current-carrying plasma inside an infinitely conducting shell. Between the plasma and the shell we insert a current sheet whose z, θ distribution is assumed to match that of the plasma perturbation. The plasma is assumed to be incompressible and to possess an equilibrium current density

$$J_z = J_0 \left(1 - (r/a_p)^\nu \right) \quad . \tag{1}$$

For simplicity we take the unperturbed z magnetic field B_z and the mass density ρ to be constant. Further, in order to relate our results to the parameter range of tokamak experiments we quantize the z wave number $k = n/R$ where R is the major radius of an equivalent torus. With these assumptions a perturbation analysis of the standard ideal MHD equations leads to a differential for the

---

[*]Research sponsored by the U. S. Atomic Energy Commission under contract with the Union Carbide Corporation.

radial displacement of a plasma element ξ of the form

$$\xi'' + P \xi' + Q \xi = 0 \tag{2}$$

where P, Q are given in the appendix. The effect of feedback enters only through the boundary condition on ξ. This is obtained from the condition of pressure continuity across the perturbed surface and can be written as

$$A \xi - B \xi' = 0 \tag{3}$$

at $r = a_p$ where A, B depend on the feedback and are given in the appendix. Equations (2) and (3) constitute an almost standard Sturm-Liouville problem. The eigenvalue parameter is the normalized growth rate γ^*. From Sturm-Liouville theory we know that when γ^* is treated as a free parameter there is a large value γ^* which corresponds to a function ξ with no nodes in the interval $r < a$. As γ^* is reduced other functions ξ with more radial nodes make their appearance. The function of Eq. (3) is merely to select the solution of the physical problem from this continuum of functions.

III. RESULTS

It is clear from Eq. (3) that feedback is effective only in determining the ratio $\xi/\xi' = B/A$. We can stabilize an unstable plasma only if we can specify this ratio such that no solution of Eq. (2) has the same value of (ξ/ξ') at $r = a_p$. The problem is illustrated in Fig. 1. It shows a plot of the ratio ξ/ξ' calculated from Eq. (2). With no feedback the ratio B/A is a line almost parallel and very close to the axis. No intersection is possible and there is no unstable eigenfunction. When q is reduced a zero crossing occurs and an instability results with a growth rate determined by the intersection point. By choosing the feedback gain functions g_1 and g_2 to be appropriate constants we can place a pole of B/A at this zero crossing to eliminate the intersection. The unstable eigenfunction is no longer an allowed solution.

The real problem becomes apparent when we attempt to take advantage of our newly won stability to reduce q still further. Figure 2 shows the ratio ξ/ξ' for a slightly smaller q. Comparing Figs. 1 and 2 one sees that lowering q below the instability threshold value results in additional modes. The number of zeros increases and the pattern shifts to the left, toward higher growth rates. These additional modes are the higher radial modes expected from Sturm-Liouville theory. Figure 3 shows the three modes corresponding to the three regions defined by the zero crossings in

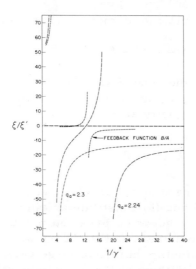

ξ/ξ'

FEEDBACK FUNCTION B/A

$q_a = 2.3$

$q_a = 2.24$

$1/\gamma^*$

Fig. 1. The ratio of the plasma's radial displacement to the slope of this displacement on the surface is plotted as a function of the normalized growth rate. The eigenfunctions of the problem are given by the intersection of these curves with the ratio specified by the boundary condition. The parameters of the calculation were $\nu = 2$, $m = 1$, $n = -1$, $a_p = 17$ cm, $a = 19$ cm, $b = 21$ cm, and $A = 7$.

Fig. 2. The same technique that stabilized the mode in Fig. 1 can still stabilize the fastest growing mode in Fig. 2. However we still have the next radial mode which has a growth rate almost equal to the mode stabilized above.

It is theoretically possible to place poles at each zero crossing and achieve complete stabilization. However from the example shown in the figures it is clear that one needs to stabilize a large number of modes to obtain a small reduction in q. In addition the placing of the feedback poles is critically dependent not only on the surface value of q but on its radial variation which affects the location of the plasma zeros. Thus to achieve stabilization one requires a complete knowledge of the plasma current distribution and a very involved feedback system. It is of course possible that the higher radial modes which peak near the plasma axis will not produce as severe plasma loss as the fundamental and thus their stabilization may not be as critical.

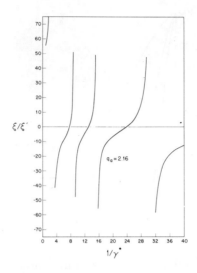

Fig. 2. The ratio of the plasma's radial displacement to the slope of this displacement on the surface is shown as a function of the normalized growth rate. The parameters used in the calculation are the same as in Fig. 1 except for a four-percent reduction in the surface q value. This illustrates the rapid accumulation of higher order radial modes when the stability limit is exceeded.

One further difficulty is the voltage and current necessary to stabilize a reactor-size plasma. We find a feedback winding impedance of

$$Z = \frac{i \pi \mu_o V_a}{2 n} \gamma^* \frac{\left(1 - (a/b)^{2m}\right) \left(1 - (^a p/a)^{2m}\right)}{1 - (^a p/b)^{2m}} \tag{4}$$

where V_a is the Alfven velocity. In the example of Fig. 1 we require feedback current constants $g_1 = -8.45$ and $g_2 = -1.22$ to stabilize the fastest growing mode. Taking γ^* to be the unstabilized growth rate this would require approximately $2 \times 10^8 \xi$ amperes at an impedance level of one ohm. Unless the normalized perturbation ξ can be detected at a level of about 10^{-4} these requirements present severe difficulties. Again one must hope that a real toroidal plasma will have slower growth rates than a cylindrical plasma for the feedback technique to be attractive.

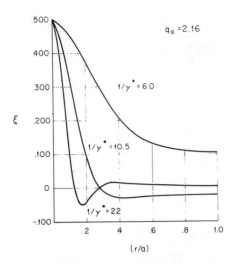

Fig. 3. Three typical eigenfunctions which can be unstable for the conditions of Fig. 2 are shown. Only the fastest growing mode can be stabilized with a simple feedback system leaving the higher order radial modes unaffected.

APPENDIX

All lengths are normalized to the plasma radius a_p and primes represent radial derivatives.

$$P = \frac{1}{r} \left[2 + \frac{m^2 A^2 - n^2 r^2}{m^2 A^2 + n^2 r^2} \right] - \left[\frac{2m(m+nq)}{q^2 H} \cdot \frac{q'}{q} \right]$$

$$Q = -\frac{n^2}{A^2} \left[1 + \frac{2 A^2}{A^2 m^2 + n^2 r^2} + \frac{2r}{H q^2} \frac{q'}{q} - \frac{4}{H^2 q^4} \right]$$

$$+ \left[\frac{4 m n^2}{q^2 H} \frac{(m+nq)}{(m^2 A^2 + n^2 r^2)} \right]$$

where m is the small azimuthal wave number, $n = k_z R$, A is the aspect ratio R/a_p, $q \equiv r B_z / R B_\theta$ and R is a length defining an equivalent toroidal periodicity.

$$H = (\gamma^*)^2 + \frac{(m + nq)^2}{q^2}$$

where γ^* is the growth rate normalized to the constant Alfven frequency $B_z/R\sqrt{\mu_0\rho}$. We assume a feedback current sheet

$$\mu_0 K_\theta = (g_1 \xi + g_2 \xi') B_z$$

where g_1 and g_2 are unspecified functions. Then the boundary condition parameters are

$$A = \left[\frac{2m}{m^2 A^2 + n^2} \frac{(m + nq)}{q^2} - \frac{H}{m^2 A^2 + n^2} \right.$$

$$- \frac{(m + nq)}{m A^2 q^2} S_a$$

$$\left. - \frac{(m + nq)}{2 nq} g_1 \left(\frac{a}{a_p}\right)^m \left(1 - \left(\frac{a_p}{b}\right)^{2m}\right) (S_a - S_b) \right]$$

where

$$S_x = \frac{\left(1 + (a_p/x)^{2m}\right)}{\left(1 - (a_p/x)^{2m}\right)}$$

$$B = \left[\frac{H}{m^2 A^2 + n^2} + \frac{(m + nq)}{2 nq} g_2 \left(\frac{a}{a_p}\right) \left(1 - \left(\frac{a_p}{b}\right)^{2m}\right) (S_a - S_b) \right]$$

and a_p, a, b are the plasma feedback winding and conducting shell radii respectively.

2.5 FEEDBACK CONTROL PROBLEMS IN TOKAMAKS

H. P. Furth and P. H. Rutherford
Princeton University
Princeton, New Jersey 08540

ABSTRACT

In the basic tokamak experiment, the copper shell is a highly
satisfactory feedback control device. For certain extensions of the
tokamak operation – for example, to compression heating – it is,
however, convenient to remove the copper shell. External feedback
control of the equilibrium position is then a natural substitute. A
more complex problem is the feedback stabilization of MHD modes.
The interchange instabilities, against which linear magnetic feedback
is ineffective, are fortunately stabilized by the minimum-average-B
property of the toroidal tokamak. Feedback stabilization of a type of
thermal instability that results in major-radius shifts is the principal
topic of this paper.

I. INTRODUCTION

The most successful feedback control element used in toroidal
confinement experiments thus far is the copper shell of the tokamak
(Artsimovich, Bobrovsky, Gorbunov, Ivanov, Kirillov, Kuznetsov,
Mirnov, Petrov, Razumova, Strelkov, Shcheglov, 1969). This
electronically simple, low-impedance, low-gain control element
performs several important functions: (1) it contributes to the estab-
lishment of plasma equilibrium in major radius; (2) it provides
stability against axisymmetric perturbations in major radius (m = 1,
n = 0 modes); (3) it contributes to stability against MHD kink modes
(principally the m = 1, 2, 3, n = 1 modes).

The copper shell, however, has serious deficiencies as a feedback control element. In regard to the provision of equilibrium (1),
it is usually too low-gain to provide the desired plasma centering,
and must be supplemented by external vertical-field circuits –
with the operation of which it tends to interfere. The response of
the copper shell at low frequencies is also unsatisfactory, because
of the skin-penetration phenomenon. In the application against major-
radius excursions (2) or kink modes (3), the copper shell has ex-
cellent frequency response; the main drawback of the copper shell as
a stabilizer is its low gain: it cannot suppress the m = 1, n = 1 mode,

and its contribution to the stabilization of the weak $m \geq 3$ modes is very small and generally superfluous (Shafranov, 1969).

At the present time there is considerable interest in replacing the tokamak copper shell with a more convenient feedback control system. One objective is to stop the interference of the shell with externally applied vertical magnetic fields, thus allowing much more positive control of the plasma major radius in the presence of runaway electrons, and also permitting strong plasma compression[*]. This application is the subject that will be discussed principally in the present paper. A more advanced objective that promises to acquire great importance in the future is the high-gain feedback stabilization of the MHD kink modes. This general subject has been discussed elsewhere (Morozov and Solovev, 1964; Arsenin, 1969) as well as in other papers in these proceedings; a few additional remarks may, however, be appropriate.

Pure MHD interchange modes evidently connot be stabilized at all be a copper shell, or by any other linear magnetic feedback scheme (Melcher, 1970) and so one must look for special toroidal configurations where the pure interchange modes are stable without feedback (Ribe and Rosenbluth, 1970). Fortunately the tokamak has just the right stability properties (Shafranov, 1969) by virtue of the minimum-average-B effect introduced by toroidal curvature: only those kink modes that bend the magnetic field lines significantly are unstable, and these can be controlled, at least in principle, by the copper shell or by some other linear feedback system.

Another convenient feature of tokamak MHD behavior, observed experimentally (Mirnov and Semenov, 1969) is that the perturbations rotate about the minor periphery of the plasma very rapidly compared with the skin penetration time of a typical copper shell. Thus it appears that the particular function of the copper shell as a kink-mode stabilizer could be taken over by a vestigial shell, only a few millimeters thick, or by some system of helical windings with equivalently short resistive relaxation times. A passive helical-winding system could be as effective as a copper shell, without interfering with the control of the vertical magnetic field; furthermore, it would lend itself to subsequent extension to active high-gain feedback control.

---

[*]H. P. Furth and S. Yoshikawa, to be published in Physics of Fluids.

II. CONTROL OF THE MAJOR RADIUS

At the high-filling-pressure end of the tokamak operating range,
one observes "anomalous" outward shifts of the plasma column
(Bobrovskii, Razumova, Shchelkov, 1968) which have been interpreted
(Furth, Rosenbluth, Rutherford and Stodiek, 1970) as relating to a
type of thermal instability. The growth of a large runaway electron
component deprives the normal plasma electron component of ohmic
heating, thus cooling it and increasing its resistivity, which leads to
a further growth of the fraction of runaway current. Due to the
centrifugal force of the runaway electrons, the thermal instability
evidently can result in shifts of the plasma major radius.

However, external control of the plasma major radius in turn
affects the runaway instability, and can even be used to suppress it.
The basic equations are:

$$j_P + j_R = j_T \propto R^{-2} , \tag{1}$$

$$j_P \eta_P = j_R \eta_R + \frac{m_e}{n_R e^2} \frac{1}{R^3} \frac{d}{dt} (j_R R^3) , \tag{2}$$

$$\frac{3}{2} n \frac{dT_e}{dt} = j_P^2 \eta_P - Q_{ie} + T_e \frac{dn}{dt} . \tag{3}$$

The subscripts P and R refer to the normal and runaway electron
components. We treat perturbations that are fast compared with the
plasma skin time: thus j_T is perturbed only due to displacement of
the major radius R. We attribute some fictitious resistivity η_R to
the runaways, so as to provide an equilibrium solution for Eqs. (1) -
(3). The variation of η_R as a result of the perturbation may well
be important, but for simplicity we will neglect it here. The electron
heat loss Q_{ie} is taken to be the classical loss to the ions (Furth,
Rosenbluth, Rutherford, and Stodiek, 1970) but the exact form of this
loss does not affect the nature of the solution. The dn/ dt term in
Eq. (3) results from the motion in R, with $n \propto R^{-2}$. Finally, we
add the feedback control equation

$$\frac{R_1}{R_o} = \frac{S}{2} \frac{j_{R1}}{j_{To}} \tag{4}$$

where the quantity S is to be prescribed by the choice of the feed-back system.

Linearizing Eqs. (1) - (3) and using (4), we obtain an instability growth rate ω given by

$$\omega \tau_H \left[1 - \frac{S \frac{j_P}{j_T}}{S + \frac{j_T}{j_R} + \omega \tau_R \left(1 + \frac{3S}{2} \frac{j_R}{j_T}\right)} \right] =$$

$$\frac{3(1 + S)}{S + \frac{j_T}{j_R} + \omega \tau_R \left(1 + \frac{3S}{2} \frac{j_R}{j_T}\right)} - \frac{3}{2} - \frac{3\alpha - 1}{2(1-\alpha)} \qquad (5)$$

(dropping the subscript o from equilibrium quantities) where $\tau_H = 3nT/ \eta_P j_P^2$ (the heating time), and $\tau_R = m_e/ n_R e^2 \eta_P$; typically $\tau_R \lesssim \tau_H$. The term in $\alpha = T_i/ T_e$ arises from the Q_{ie} term in Eq. (3). Analyzing Eq. (5) for $\tau_R \ll \tau_H$ by dropping the term in $\omega \tau_R$ we obtain a stability condition

$$\left[\frac{3}{2} + \frac{3\alpha - 1}{2(1-\alpha)} - \frac{3(1 + S)j_R}{j_T + S j_R} \right] \left[\frac{S + j_T/ j_R}{S + (j_T/ j_R)^2} \right] > 0 \qquad (6)$$

As we shall see below, if the copper shell dominates the equi-librium in R, then S is typically in the range 1 >> S > 0. The perturbation is then generally unstable for $j_R \gtrsim j_P$, the charact-eristic growth time being the heating time τ_H (if $\tau_R \lesssim \tau_H$). For larger positive values of S, the mode would become unstable even at low levels of j_R/ j_P. If, by means of suitable feedback, we could make S somewhat negative, this would be favorable to stabilization: clearly S = -1, for example, would always give stability. Large negative values of S are once more destabilizing: indeed instability always occurs if $S < - (j_T/ j_R)^2$. Equation (6) would predict stability whenever $- (j_T/ j_R)^2 < S < -j_T/ j_R$; in this range, however, there are faster growing instabilities with $\omega \sim \tau_R^{-1} >> \tau_H^{-1}$. For these, Eq. (5) gives

$$\omega \tau_R = - \frac{S + (j_T/j_R)^2}{3S/2 + j_T/j_R} \tag{7}$$

so that instability occurs whenever $-(j_T/j_R)^2 < S < -2j_T/3j_R$.

To illustrate the magnitude of the feedback parameter S in several "realistic" major-radius positioning systems, we will use a simple model:

$$\frac{2R(R-R_c)}{b^2} = \log \frac{b}{a} + \frac{4\pi n_R m_e V_R^2}{B_\theta^2} - \frac{B_\perp}{I_T} R, \tag{8}$$

where we follow Shafranov (1963) but neglect the terms involving ordinary plasma pressure and internal plasma inductance and add a term involving the centrifugal force of the runaways. The total plasma current is I_T, the center of the copper shell is at $R = R_c$, its minor radius is b, and the plasma radius is a. The externally controlled vertical magnetic field is $B_\perp$, and we will assume for it the feedback prescription

$$\frac{B_{\perp 1}}{B_{\perp o}} = g \frac{R_1}{R_o} \tag{9}$$

(which includes the effect of a spatial R-dependence of $B_\perp$).

If we have $b^2/R_c^2 \ll 1$, then the copper shell dominates in Eq. (8), and we obtain, using Eq. (4),

$$O < S = 2 \frac{b^2}{R_c^2} \frac{j_R}{j_T} N \ll 1, \tag{10}$$

where $N = \pi a^2 n_R e^2/m_e c^2$ is typically a number not too far from unity.

Going to the opposite extreme, if we remove the copper shell and assume a case where the centrifugal force and $B_\perp$ terms dominate in Eq. (8), we obtain

$$S = \frac{4}{g - 3} \frac{j_T}{j_R} . \tag{11}$$

If g is made large and positive, the result is similar to that obtained in the case of the copper shell: S becomes small and positive. If $-3 < g < 3$, then $S < -2 j_T / 3 j_R$, which has been shown always to lead to instability, however small j_R: indeed, if $-3 < g < 3 - 4 j_R / j_T$, the instability grows on the faster time scale τ_R. For $g \to -3$, the growth rate in (7) becomes large, so that the neglect of the inertial term $d^2 R / dt^2$ in (8) becomes important; finally, for $g < -3$, this term gives an MHD-like (Yoshikawa, 1964) mode growing on the dynamic time scale. The most reliable approach would therefore appear to be a very "stiff" feedback, in the sense $g \gg 1$, though even this approach does not insure stability against the runaway mode once the condition $j_R \gtrsim j_P$ has been reached.

In view of the extensive idealizations adopted in the present model, an attempt to give detailed prescriptions for experimental feedback stabilization would evidently be unrealistic. The present discussion is meant only to outline the nature of the problem and suggest the importance of further experimental and theoretical work.

ACKNOWLEDGMENT

We should like to thank Drs. M. N. Rosenbluth and W. Stodiek for valuable discussions.
"This work was performed under the auspices of the U.S. Atomic Energy Commission, Contract No. AT(30-1)-1238."

2.6 FEEDBACK STABILIZATION OF A HIGH-β, SHARP-BOUNDARIED PLASMA COLUMN WITH HELICAL FIELDS

F. L. Ribe

University of California, Los Alamos Scientific Laboratory
Los Alamos, New Mexico 87544

and

M. N. Rosenbluth

Institute for Advanced Study, Princeton, New Jersey 08540

ABSTRACT

A sharp-boundaried high-β Stellarator of ℓ-fold helical symmetry is MHD unstable, particularly to the $m = 1$, long-wavelength mode. For this mode the instability is driven by $\ell \pm 1$ fields introduced by the unstable displacement. Compensating fields (or, equivalently, plasma distortions) to counteract the instability can be introduced externally and controlled by a feedback arrangement. The self-adjointness property of the MHD energy-principle operator is used to derive the relationship between the external fields and the unstable displacement.

I. INTRODUCTION

Recently the equilibrium (Blank, Grad, and Weitzner, 1969; Rosenbluth, Johnson, Greene, and Weimer, 1969) and stability (Rosenbluth, Johnson, Greene, and Weimer, 1969; Grad and Weitzner, 1969) of a sharp-boundaried theta-pinch plasma column of major radius of curvature R (minor radius a) have been studied in the presence of helical fields whose scalar potential is of the form

$$\chi = (B^{(o)}/h) \left[C_\ell I_\ell (hr) + D_\ell K_\ell (hr) \right] \sin(\ell\theta - hz).   \qquad (1)$$

The zero-order field $B^{(o)}$ is directed along z and has the values B_o and $\sqrt{1 - \beta}\, B_o$ outside and inside the plasma column. The first-order excursion δ_ℓ of the plasma radius about $r = a$ is given by

$$r = a + a \delta_\ell \cos(\ell\theta - hz)   \qquad . \qquad (2)$$

To second order the effect of major curvature is not observable, and a straight plasma column subject to a helical field is unstable to displacements given in lowest order by $\xi(m, k) = \xi_o \cos(m\theta - kz)$. Following recent experimental results on the $\ell = 0$ system (Bodin, McCarten, Newton, and Wolf, 1969; Little, Newton, Quinn, and Ribe, 1969; Fuenfer, Junker, Kaufmann, Neuhauser, and Seidel, 1968) which show the presence only of the m = 1 mode, we limit consideration to m = 1. The instability growth rate γ_ℓ is given by (Rosenbluth, Johnson, Greene, and Weimer, 1969)

$$\gamma_\ell^2 = \tfrac{1}{2}\beta G_\ell \delta_\ell^2 (hv_A)^2 - (2-\beta)(kv_A)^2 , \qquad (3)$$

where $v_A = B_o/(4\pi\rho)^{\frac{1}{2}}$ is an "Alfven" speed, ρ is the plasma density, and

$$G_o(\beta, ha) = 2(1-\beta)(3-2\beta)/(2-\beta) \qquad (4)$$

$$G_1 = [(2-\beta)(4-3\beta)/4(1-\beta)](ha)^2 \qquad (5)$$

$$G_\ell = (2-\beta)(\ell-1) \qquad \ell > 1 , \qquad\qquad ha \ll 1. \qquad (6)$$

The case $\ell = 1$ has an especially low growth rate, owing to the presence of the small factor $(ha)^2$ in Eq. (5). In the Scyllac ordering (Grad and Weitzner, 1969) under consideration here, these growth rates also apply in the presence of curvature.

II. CONTROL OF EQUILIBRIUM AND STABILITY BY MEANS OF COMBINATIONS OF HELICAL FIELDS

The energy-principle calculation (Rosenbluth, Johnson, Greene, and Weimer, 1969) shows that a first-order m = 1, k $\approx$ 0 displacement ξ_o gives rise to ($\ell \pm 1$) helical distortions (or, equivalently, magnetic fields) of the plasma column in next higher order which account for the instability.

Such distortions can also be produced by second-order $\ell \pm 1$ external fields of strength $C_{\ell\pm1}^{ext}$. The effect of the $\delta_{\ell\pm1}^{ext}$ is to give an "interference" of the ℓ and $\ell \pm 1$ components, producing an asymmetrical distortion of the plasma column. In the direction of largest distortion there is a force per unit length (Ribe, 1969) given by

$$F_{\ell,\ell\pm1}^{ext} = - (\beta/8)B_o^2 h^2 a^3 \delta_\ell \delta_{\ell\pm1}^{ext} \left\{ 1 + (1-\beta)\left[1 + \frac{\ell(\ell+1)}{(ha)^2} \right] \frac{I_\ell I_{\ell\pm1}}{I'_\ell I'_{\ell\pm1}} \right\}, \quad (7)$$

where the Bessel functions and their derivatives are evaluated at $r = a$.

Corresponding to the growth rate of Eq. (3), one can write a destabilizing force (Ribe, 1969)

$$F_\ell = (\beta/8)B_o^2 (ha)^2 G_\ell \delta_\ell^2 \xi_o, \quad (8)$$

which can be compensated by $F_{\ell,\ell\pm1}$, provided for example, the $\delta_{\ell\pm1}^{ext}$ is made proportional to ξ_o by feedback.

III. DERIVATION OF THE FEEDBACK RELATIONSHIP FROM THE ENERGY-PRINCIPLE EULER EQUATIONS

The force $F_{\ell,\ell\pm1}^{ext}$ of Eq. (7) was derived (Ribe, 1969) by evaluating the increment $\langle B_\ell \cdot B_{\ell\pm1}^{ext} \rangle /4\pi$ of magnetic pressure across the flux surface separating the plasma and vacuum in direction of its inward normal, multiplying by $\cos\theta$, and integrating over the interface. The self-adjointness property of the force operator

$$f = \rho\gamma^2 \xi = j_{1p} \times B_e + j_e \times B_1 - \nabla p_1 \quad (9)$$

of the MHD energy principle (Bernstein, Frieman, Kruskal, and Kulsrud, 1958) can be used to show the general form of the interaction of external fields with the plasma, of which $F_{\ell,\ell\pm1}$ is a special case. In Eq. (9) the subscript e refers to the equilibrium field of Eq. (1) and B_1 to the field induced by the perturbing plasma displacement ξ:

$$B_1 = \nabla \times (\xi \times B_e) . \quad (10)$$

The plasma current density is given by

$$j_{1p} = (\nabla \times B_1)/4\pi - j^{ext}, \quad (11)$$

where j^{ext} is the plasma current to be controlled by feedback. Let subscripts I and II refer to the situations without and with feedback:

$$\rho\gamma_I^2 \xi_I = \mathcal{L} \xi_I \quad (12)$$

$$\rho\gamma_{II}^2 \xi_{II} = \mathcal{L}\xi_{II} - j^{ext} \times B_e . \quad (13)$$

Taking scalar products of Eq. (12) with ξ_{II} and Eq. (13) with ξ_I, subtracting, integrating over volume, and using the self-adjointness property of $\mathcal{L}$ we obtain

$$(\gamma_I^2 - \gamma_{II}^2)\int \rho \xi_I \cdot \xi_{II} \, d\tau = - \int j^{ext} \cdot \xi_I \times B_e \, d\tau \, , \qquad (14)$$

where in perturbation theory ξ_{II} in the left-hand side integral may be taken to be approximately equal to the displacement ξ_I in the absence of feedback. Substituting $j^{ext} = (\nabla \times B^{ext})/4\pi$, the right-hand side of Eq. (14) can be integrated by parts over all space:

$$-\int \left(\nabla \times B^{ext}\right) \cdot \xi_I \times B_e \, d\tau = - \int \nabla \cdot \left[B^{ext} \times (\xi_I \times B_e) \right] d\tau$$

$$- \int B^{ext} \cdot \nabla \times (\xi_I \times B_e) d\tau = - \int dS \cdot \xi_I \left\langle B_e \cdot B^{ext} \right\rangle$$

$$- \int B^{ext} \cdot \nabla \times (\xi_I \times B_e) d\tau \, , \qquad (15)$$

where the surface integral is over the flux surface forming the plasma-vacuum interface. However, $\nabla \times (\xi_I \times B_e) = \nabla\chi$, since the perturbed field is curl-free both in the vacuum and plasma regions. Hence the volume integral of Eq. (15) reduces to

$$\int B^{ext} \cdot \nabla\chi \, d\tau = \int \nabla \cdot \left(B^{ext} \chi \right) d\tau = \int dS \cdot \left\langle B^{ext} \chi \right\rangle \, . \qquad (16)$$

For neutral stability ($\gamma_{II} = 0$) Eq. (14) thus reduces to

$$\gamma_I^2 \int \rho \, \xi_I^2 \, d\tau \approx - (4\pi)^{-1} \left\{ \int dS \cdot \xi_I \left\langle B_e \cdot B^{ext} \right\rangle + \int dS \cdot \left\langle B^{ext} \chi \right\rangle \right\} \, . \qquad (17)$$

In order to apply these general results we now utilize the properties of the equilibrium fields of Eq. (1), as well as those components $\chi_{\ell\pm1}^{(2)}$ of the potential which arise during the unstable modes without feedback. Here also pressure balance must apply for the $\ell \pm 1$ Fourier component. Now, however, there is an additional force because the m = 1 displacement ξ_o must move against the gradient of unperturbed field pressure. We substitute the $\chi_\ell^{(2)}$ and $\chi_{\ell\pm1}^{(2)}$ fields into Eq. (17) and take account of the jump in zero order field. Averaging over θ then gives the desired result:

$$\gamma_I^2 \int \rho \xi_I^2 d\theta = \frac{\beta}{8} B_o^2 h^2 a^3 \delta_\ell \delta_{\ell\pm1}^{ext} \xi_o \left\{ 1 + (1-\beta)\left[1 + \frac{\ell(\ell \pm 1)}{h^2 a^2} \right] \frac{I_\ell I_{\ell\pm1}}{I_\ell' I_{\ell\pm1}'} \right\} \, . \qquad (18)$$

Interpreted as a force, the right-hand side of Eq. (18) evidently yields Eq. (7).

CHAPTER 3

FEEDBACK EXPERIMENTS IN Q-DEVICES,
DISCHARGES, AND SOLIDS

3.1 NONLINEAR COLLISIONLESS INTERACTION BETWEEN
ELECTRON AND ION MODES AND FEEDBACK
STABILIZATION IN FUSION PLASMAS

A. Y. Wong, D. R. Baker, and N. Booth
Department of Physics, University of California,
Los Angeles, California 90024

ABSTRACT

An efficient method of interaction with ion modes in fusion and space plasmas through the excitation of resonant electron modes is proposed. The feasibility has been demonstrated in laboratory experiments over a wide density range (10^8-10^{10}/cm^3) and a wide ion frequency range (1 kHz - 500 kHz). The collisionless nature of the nonlinear interaction between electron and ion modes is emphasized leading to the feasibility of feedback stabilization in hot collisionless fusion plasmas. Measurements of nonlinear coupling coefficients are presented which support the theoretical mode-mode coupling model.

I. INTRODUCTION

We are presenting an efficient method of interaction with ion modes without probes, by means of collisionless coupling between resonant electron modes. The observation of the effect has been reported (Wong, Baker, and Booth, 1970) over a wide density range (10^8-10^{11}/cm^3) and a wide ion frequency range (1 kHz - 500 kHz). Here we are discussing the development of a simple physical model and preliminary data which seems to favor this model. This work demonatrates the feasibility of interacting with a fusion plasma without the use of probes.

We are concerned with electron and ion modes that propagate nearly perpendicular to the magnetic field. A simple representation of these modes is shown in Fig. 1. Coupling among electron modes has been demonstrated by Ellis and Porkolab (1968) and the ion modes by the UCLA group (Wong, Goldman, Hai, and Rowberg, 1968). We shall mainly dwell on the intercoupling between electron and ion modes in an inhomogeneous plasma as is usually encountered. These two kinds of modes are similar in several aspects; they propagate nearly perpendicular to B, have long axial wavelength, and derive their resonant characters from the combined cyclotron and collective effects. In the case of drift waves the density and temperature gradients supply the free energy and the azimuthal periodicity make the modes discrete and resonant. These modes are easily controlled by microwave horns, inserted conveniently through the side ports as shown in Fig. 2, which radiate at the upper hybrid resonance.

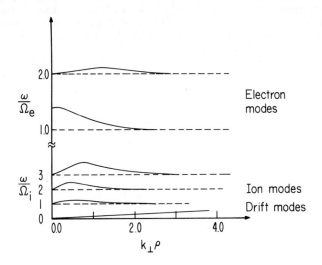

Fig. 1. Dispersion characteristics of electron and ion modes.

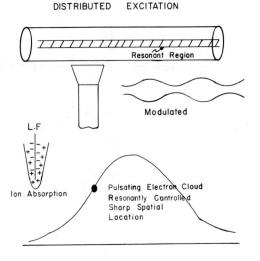

Fig. 2. Schematic of distributed excitation.

The density gradient gives rise to a radially sharply localized reso-
nant region since $\omega_{UH}^2 = \omega_p^2(r) + \omega_c^2$. Either a low-frequency modula-

tion of the high-frequency mode at the ion resonance or the injection of two microwave beams at ω_1 and ω_2 separated by ω_i gives an enhanced ion response. One can view the coupling as due to an electron cloud resonantly controlled, pulsating at the ion frequency ω_i. The pulsation results from a nonlinear coupling of the input frequencies. The coupling to the ions takes place by a direct electric field interaction. Because the electron mode has high phase velocity along B the resonant region is distributed and acts effectively as a distributed source to couple to ion modes over the entire length; the high frequency employed permits good focussing of the incident beam. In contrast the direct excitation of the ion mode by a probe is localized and encounters shielding by the sheath and ions are lost to the probe. Effectively, we have made use of the electrons that normally shield the probe to become a source of field or a virtual distributed probe.

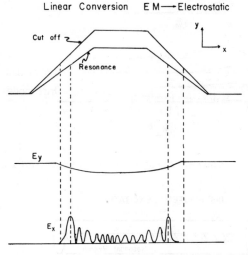

Fig. 3.   Linear conversion from electromagnetic to electrostatic waves.

Figure 3 shows an interesting characteristic of the electron resonant mode at ω_{UH}. As the electromagnetic wave propagates through the density gradient its electric vector E_y rotates into E_x, in alignment with the propagation vector. This makes it efficient to excite the electrostatic Bernstein modes (Fig. 1) propagating along x. The result of an extensive Vlasov calculation including the full set of Maxwell equations in an inhomogeneous plasma by L. Hedrick (1969) is represented in Fig. 3. As confirmed experimentally, there is a sharp electric field gradient of the order of a few electron cyclotron Larmor radii. This is an important result which explains the localization of the coupling and the nonlinearity responsible for the coupling. The square of the electric field measured at the localized region is enhanced typically by a factor of 10 due to plasma resonant effects at ω_{UH}. The evanescant region that normally shields the resonance (Fig. 3) can be overcome by adjusting plasma parameters, the magnetic field and density such that the free space wavelength of the upper hybrid frequency is large compared to the evanescent region.

The mode coupling process is completely reversible making it possible to scatter the incident microwave beam (less than 1 mW) by an ion wave. The main emphasis here is that a completely probe-less detection and excitation system is possible laying the way for our probeless feedback scheme.

II. EXPERIMENT

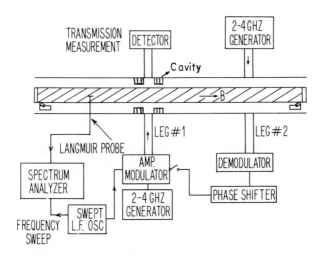

Fig. 4. Experimental arrangement showing two S-band systems and a Langmuir probe. The amplitude modulator and the spectrum analyzer are controlled by the same low-frequency swept oscillator such that the spectrum analyzer always tracks the frequency of the ion wave being excited.

The actual experimental arrangement is indicated in Fig. 4. Two sets of S band systems, one for the excitation and the other for detection through forward scattering and an RF Langmuir probe are shown with a Q-machine plasma which is highly ionized and can be operated in regimes where electron-neutral and electron-ion collisiona are unimportant. The amplitude modulator and the spectrum analyzer are controlled by the same low frequency swept oscillator such that the spectrum analyzer always tracks the frequency of the ion wave being excited. Figure 5 shows the response of the spectrum analyzer as the modulation frequency is swept. Besides drift waves and edge oscillations one observed electrostatic ion cyclotron waves and harmonics up to the 7th ($\approx$ 350 kHz).

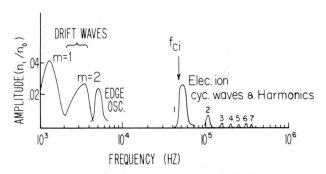

Fig. 5. Amplitude response of the plasma monitored either by a negatively biased probe or from an S-band detection system (leg No. 2) versus the modulation frequency on the excitation S band (leg No. 1); n $\approx 10^9$ cm^{-3}, B = 1.25 kG. Density gradient waves of azimuthal numbers m = 1, 2, an edge oscillation, and electrostatic ion cyclotron waves up to the 7th harmonic are shown.

We have operated in density regimes close to 10^9 cm^{-3} such that $\omega_{UH}\tau_{ee} \gg 1$, $\omega_i\tau_{ee} \gtrsim 1$ and $\omega_i\tau_{ei} \gg 1$, where τ_{ei} is the electron-ion energy exchange time and the electron mean free path is of the order of the length of the machine. We have operated at a high modulation frequency to demonstrate that even when under conditions where the low thermalization rate does not permit the electron temperature to fluctuate at the ion frequency, coupling is still feasible.

As shown in Fig. 6 the density-gradient drift waves which are less damped near the center are excitable by a μw frequency at 3.74 GHz compared to a μw frequency of 3.71 GHz for the edge oscillations which are more excitable at the edge where the temperature gradient is large. This is consistent with our picture of the upper hybrid mode which has a lower frequency at the edge.

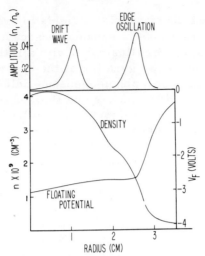

Fig. 6.   Demonstration of radial selectivity of excitation of ion resonances through adjustment of incident microwave frequency. The radial positions of excited ion resonances (upper trace) are shown with reference to the density and potential profiles (lower trace). f_{ce} = 3.70 ± 0.005 GHz.

III.   THEORY AND DISCUSSION

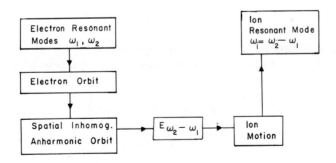

Fig. 7.   Block diagram of elements in coupling mechanism.

The collisionless coupling can be explained by a block diagram (Fig. 7).   The incident EM wave is converted to an electrostatic resonant mode.   The electron orbit becomes larger as resonant acceleration takes place.   Spatial inhomogeneity gives rise to an anharmonic orbit coupling the two incident fields to produce $\omega_2 - \omega_1$ and $\omega_2 - \omega_1$. The ion response is enhanced whenever $\omega_i \approx \omega_2 - \omega_1$. The functional dependence can be illustrated with a rather simple model considering the electron as gyrating at the upper hybrid frequency in the presence of a highly inhomogeneous field. The model is similar to the one proposed by Kuckes and Dawson (1965) to explain the anomalous cyclotron emission. In the following we shall analyze a particular case of coupling, that between the high-frequency electron mode and the low-frequency electrostatic ion cyclotron mode propagating along the density gradient.

Consider the motion of electrons under the influence of an external magnetic field $\overrightarrow{B_0}$, an external oscillating electric field E_e, and an internal self-consistent electric field E_i governed by Poisson's equation. We shall adopt a slab geometry with the magnetic field in the z-direction, and the density gradient and electric fields in the x-direction. The position of an electron is written as the position of the guiding center $\overrightarrow{r_0}$ plus the small excursion from this position:

$$\overrightarrow{r} = \overrightarrow{r_0} + \delta_x(\overrightarrow{r_0}, t)\,\hat{x} + \delta_y(\overrightarrow{r_0}, t)\,\hat{y}$$

The equations of motion are

$$\ddot{\delta}_x = -\frac{e}{m}(E_e + E_i) - \Omega_e \dot{\delta}_y \tag{1}$$

$$\ddot{\delta}_y = \Omega_e \dot{\delta}_x \tag{2}$$

where Ω_e is the electron cyclotron frequency.
The internal electric field is given by Maxwell's equation ($\nabla \times B = 0$)

$$\dot{E}_i = -4\pi j = 4\pi n_0 e \dot{\delta}_x \quad \text{or} \quad E_i = 4\pi n_0 e \delta_x \tag{3}$$

Combining these equations, the displacement δ_x can be written as a harmonic oscillator equation driven by the external electric field.

$$\ddot{\delta}_x + \omega_H^2(x)\delta_x = -\frac{e}{m}E_e \tag{4}$$

where

$$\omega_H^2(x) = \omega_p^2(x) + \Omega_e^2 = \omega_H^2(x_0) + \omega_p^2(x_0)\frac{1}{n_0}\frac{\partial n_0}{\partial x}\delta_x$$

is the upper hybrid frequency. Keeping only the first term in this expansion and calculating the response to two applied frequencies, ω_1 and ω_2, we obtain:

$$\delta_x = -\frac{e}{m}\left[\frac{E_{e1}}{-\omega_1^2 + \omega_H^2}(e^{-i\omega_1 t} + c.c.) + \frac{E_{e2}}{-\omega_2^2 + \omega_H^2}(e^{i\omega_2 t} + c.c)\right] \quad (5)$$

Inserting this result into Eq. (4) in the spirit of perturbation theory, looking at the response at $\Delta\omega = \omega_1 - \omega_2$, and using $\Delta\omega \ll \omega_1$, ω_2, ω_H one obtains

$$\delta_x(\Delta\omega) = \frac{-\omega_p^2 K e^2}{\omega_H^2 m^2}\left[\frac{E_{e1}E_{e2}\,e^{i(\omega_1-\omega_2)t}}{(\omega_1^2 - \omega_H^2)(\omega_2^2 - \omega_H^2)} + (1 \longleftrightarrow 2)\right] \quad (6)$$

where $K = (1/n_0)\,\partial n_0/\partial x$. Thus the electrons oscillate in the x direction at the difference frequency $\Delta\omega$. Due to the fact that there is a density gradient this electron motion causes charge separation. Inserting this charge separation into the usual dynamics for the electrostatic ion cyclotron wave, we obtain

$$n_e = n_0 + \frac{\partial n_0}{\partial x}\delta_x(\Delta\omega) + n_0\,\psi(\Delta\omega)$$

$$n_i = n_0 + n_0\left(\frac{k_x^2 a_i^2}{(\Delta\omega)^2 - \Omega_i^2 - k_x^2 a_i^2}\right)\psi(\Delta\omega) \quad (7)$$

where $a_i^2 = kT/m_i$ and $\psi(\Delta\omega) = e\,\phi(\Delta\omega)/kT$.
Inserting this into Poisson's equation we obtain

$$\varepsilon_i\,\psi(\Delta\omega) = -k_p^2 K\delta(\Delta\omega) \quad (8)$$

where $k_D^2 = \frac{4\pi n_0 e^2}{kT}$ and $\varepsilon_i = k^2 + k_D^2\left(\frac{k_x^2 a_i^2}{(\Delta\omega)^2 - \Omega_i^2 - k_x^2 a_i^2} - 1\right)$.

Equation (8) can be rewritten in the form

$$\psi(\Delta\omega) = M E_{e1} E_{e2}$$

where

$$M = \frac{k_D^2 K^2 \omega_p^2 e^2}{m_e^2 \omega_H^2 \varepsilon_i (\omega_o^2 - \omega_H^2)^2} \quad , \quad \omega_o \approx \omega_1, \ \omega_2$$

By taking into account the density dependence of the dielectric constants we arrive at the following limits of the coupling coefficients:

$$M \propto n_o^2 K^2 \text{ for } k << k_D; \text{ and } M \propto n_o^2 K^2 \text{ for } k \approx k_D.$$

The approximate experimental confirmation of the functional variation of the coupling coefficient M was made by measuring the ion response n/n_o as a function of the input microwave power (Fig. 8) over a density from 10^9 to 10^{10} . The coupling coefficient proportional to the slopes of the curves varies approximately as n^2 (Fig. 9).

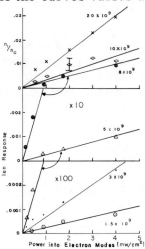

Fig. 8. Ion response vs power into electron modes.

We have chosen the collisionless mode coupling model instead of a macroscopic model such as the temperature fluctuation model proposed by Hendel, Chu, Perkins, and Simonen (1970) for three reasons; first, the high frequency $\omega_2 + \omega_1$ is observed, which is not predicted by the temperature model; second, by sampling the electron temperature after the turning on of a microwave pulse we find that T_e undergoes changes in the order of 1 msec, a rate which is too slow to account for the rapid ion oscillations that are observed. In fact, the stronger the gain in energy from the resonant interaction, the smaller is the collisional rate; third, a single sideband excitation instead of the amplitude modulation also excites the ion resonance. Indeed, we have found that a single microwave frequency has sufficient bandwidth to couple to the ion mode and modulation is not necessary if one is content with a larger threshold (a factor of 3).

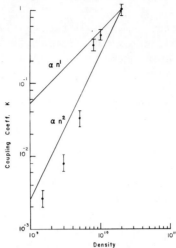

Fig. 9.   Normalized coupling coefficient vs. density.

The application of our technique to feedback stabilization of a m = 2 drift mode in the collisionless regime is shown in Fig. 10. The effect of the microwave in providing the positive and negative feedback is evident.   Our techniques also permit the feedback stabilization or detection of unstable electrostatic ion cyclotron waves such as encountered in mirror machines.

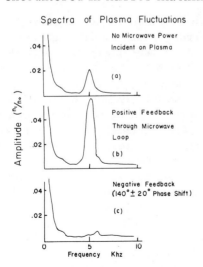

Fig. 10.   Demonstration of feedback stabilization of m = 2 drift wave.   n = $5 \times 10^9 / cm^3$.

3.2 REMOTE FEEDBACK STABILIZATION OF DRIFT INSTABILITY BY MICROWAVES

T. K. Chu, H. W. Hendel, F. W. Perkins, and T. C. Simonen
Princeton University
Princeton, New Jersey 08540

ABSTRACT

A modulated microwave source is used in remote feedback stabilization of the collisional drift instability in a Q-machine cesium plasma at $n \simeq 5 \times 10^{10} cm^{-3}$. The microwave heats electrons by resonance absorption at interior plasma locations selected by adjusting the microwave frequency to the local upper-hybrid frequency $[f_{ce}^2 + f_{pe}^2]^{1/2} \simeq 11 GHz$. Penetration of the evanescent exterior plasma region is achieved by introducing a weak magnetic field gradient ($\nabla B/ B \approx 0.03 cm^{-1} \ll \nabla n/ n \approx 1 cm^{-1}$) with an iron block placed outside the plasma column. The modulation phase (< 180°) and microwave amplitude ($\simeq 0.1$ mW) necessary for stabilization agree qualitatively with a linear theory including a modulated heat source. Upon stabilization, improved plasma confinement is measured.

I. INTRODUCTION

We report results of work on remote feedback suppression of the drift instability in the strongly collisional regime by feedback modulation of microwaves resonantly interacting with the plasma at the upper hybrid frequency (Hendel, Chu, Perkins, and Simonen, 1970). For optimum stabilization the phase of the modulated microwave energy source lags approximately 180° behind that of the drift wave, in qualitative agreement with a linear-theory interpretation based on modulation of an energy source, and hence electron temperature; the modulation in effect counteracts the destabilizing expansion of electron flow along the field. Upon stabilization plasma density increases concomitantly. In Section II, a linear stability theory for collisional drift waves including a modulated energy source is presented. In Section III the experiment is described. The experimental results in terms of feedback phase delay and gain and their comparisons with theoretical predictions of linear growth rates are given. Finally the significance of the experiment is discussed and comparisons of different interpretations are made.

II. THEORY

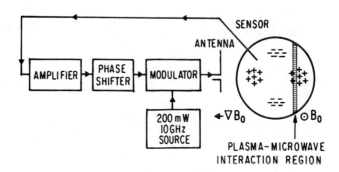

Fig. 1. Schematic of remote feedback stabilization by modulated microwave beam. The plasma and microwave interact at a region where the upper hybrid frequency equals the microwave frequency. The magnetic field gradient is produced by an iron block (not shown) opposite to the microwave antenna.

The simple feedback scheme used is shown in Fig. 1. For microwaves at frequency f_μ, the plasma-microwave interaction occurs at a plasma region where the local upper hybrid frequency equals the microwave frequency. Localized heating results (Stix, 1960; Wong, 1965; Wong and Kuckes, 1964). In a collisional plasma, low-frequency modulation of the microwave therefore results in modulation of the electron temperature, since $ft \ll 1$, where $f \approx 2$ kHz is the drift-wave frequency and $t \approx 5 \times 10^{-5}$ sec is the electron thermal relaxation time. This fast heating of the plasma and the fast thermal diffusion are also confirmed in recent experiments by Stefanov (1970). The stability theory can therefore be carried out by including an energy source term S_e in the electron energy equation.[*] The relevant electron-fluid equations, in standard notation, are: (Braginskii, 1965)

$$0 = -\nabla_\perp \tilde{P}_e - en_o(-\nabla_\perp \tilde{\phi} + \tilde{\underset{\sim}{u}}_{e\perp} \times B) - e\tilde{n}(\underset{\sim}{v}_d \times \underset{\sim}{B}) \tag{1}$$

$$0 = -\nabla_{||} \tilde{p} + en_o\nabla_{||}\tilde{\phi} - c_r m_e n_o \nu_{ei} \tilde{u}_{e||} - c_t n_o \nabla_{||}\tilde{T} \tag{2}$$

$$\partial \tilde{n}/\partial t + \tilde{\underset{\sim}{u}}_{e\perp}\cdot\nabla n_o + v_d \partial\tilde{n}/\partial y + n_o\nabla\cdot\underset{\sim}{u}_e = 0 \tag{3}$$

$$(3/2)n_o(d\tilde{T}/dt) - T_o(d\tilde{n}/dt) + \nabla\cdot\tilde{q} = Q\tilde{n} k T_o \tag{4}$$

where $\tilde{q} \equiv -c_\kappa(n_o T_o/m_e\nu_{ei})\nabla_{||}\tilde{T} + c_{||}n_o T_o\tilde{u}_{e||} - (5/2)(n_o T_o/m_e\Omega_e)$. $(\hat{z}\times\nabla\tilde{T})$, $c_\kappa = 0.51$, $c_{||} = 0.71$, $c_\kappa = 3.16$, and $\tilde{S} = Q n k T$. Combining Eqs. (1)-(4), we obtain: (Hendel, Chu, and Politzer, 1968; Tsai, Perkins, and Stix, 1970)

$$\frac{\tilde{n}}{n_o} = \frac{(1/t_{||} - i\omega_e)+(2/3)(\omega_e/\omega t_{||})[\,(1+c_t)^2 + c_r c_e(1 + i/\omega_e t_{||})]}{(1/t_{||}-i\omega)+(2/3t_{||})[\,(1+c_t)^2 + c_e c_r(1+i/\omega t_{||}) - i(1+c_t)\,Q/\omega]}\frac{e\tilde{\phi}}{KT} \tag{5}$$

where $1/t_{||} = k_{||}^2 KT/c_r m_e\nu_{ei}$, $\omega_e = (-k_y KT/eB)(\nabla n_o/n_o)$. The corresponding ion-fluid equation is: (Hendel, Chu, and Politzer, 1968)

$$\frac{\tilde{n}}{n} = \frac{i(-b\omega + \omega_e) + 1/t_\perp}{i\omega(1+b) - 1/t_\perp}\frac{e\tilde{\phi}}{kT} \tag{6}$$

where $b = 1/2\, k_\perp^2\nu_i^2$, $1/t_\perp = (1/4)b^2\nu_{ii}$. The linear dispersion relation is obtained by combining Eqs. (5) and (6).

---

[*]We use a uniformly distributed source whose strength is proportional to the local instability amplitude. For feedback consisting of local sources only, by constructing a quadratic form consisting of positive definite integrals, it can be shown that the present results retain the basic features.

For the limiting cases, optimum stabilization occurs when the energy source lags behind the wave from 180° ($\omega t_{\parallel} \gg 1$) to 270° ($\omega t_{\parallel} \ll 1$).

III. EXPERIMENT

The experiment was performed in the cesium plasma of the Q-1 device ($n_e \simeq 5 \times 10^{10}$ cm^{-3}, T $\simeq 2800°$ K, B $\simeq$ 4kG), with a Langmuir probe as detector and a half-wave dipole antenna (located ~2 cm outside the plasma column, at its mid-plane) as suppressor. The amplified and phase-shifted ion-saturation-current instability signal modulates the microwave source so that its output power is proportional to the instability amplitude. Penetration of the evanescent exterior plasma region is achieved by placing a 2 × 10 × 15⋯ cm (in r, θ, z directions) iron block, 5 cm off axis, diametrically across from the antenna to introduce a weak magnetic field gradient ($\nabla B / B \approx 0.03 \ll$ $| \nabla n_o / n_o | \approx 1$) in the otherwise constant and uniform field.

Localized plasma heating is detected by steady-state probe-measurements of electron temperature and of the local change of plasma potential. Figure 2 shows the shift of the microwave-plasma

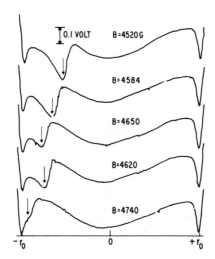

Fig. 2. Shift of the microwave-plasma interaction region as indicated by the radial distribution of plasma potential. The magnetic field is varied and the microwave frequency is kept constant. Cesium plasma, T $\approx 2800°$ K, $n_o \simeq 5 \times 10^{10}$ cm^{-3} .

interaction region, as indicated by the radial distribution of plasma potential, when the magnetic field is varied and the microwave frequency is kept constant. Figure 3 shows the radial location of the interaction region when the microwave frequency is varied and the

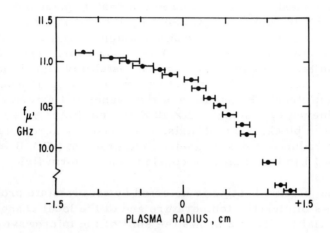

Fig. 3. Radial location of the microwave-plasma interaction region. The microwave frequency is varied and the magnetic field is kept constant. Cesium plasma, $T \simeq 2800°$ K, B = 4200G, $n_o = 5 \times 10^{10}$ cm^{-3}.

magnetic field is constant. (The monotonic variation across the plasma diameter is due to the monotonic variation of the magnetic field produced by the iron block placed at approximately r = -6.5cm). In these measurements, the input microwave power is nearly 100 times that used in instability suppression. The interaction region extends 1 - 3 mm, much less than the wavelength or density gradient scale length. At reduced microwave power, the instability region becomes even more localized. These measurements are in agreement with the radial location determined by the upper hybrid frequency, $\omega_h^2 = \omega_{ce}^2 + \omega_{pe}^2$.

IV. RESULTS

Figure 4 shows the instability amplitude, change of plasma density at the axis, and instability frequency as a function of microwave feedback phase delay. Also shown in Fig. 4a is the calculated linear growth rate of the instability according to the linear dispersion

relation obtained from Eqs. (5) and (6) for the experimental conditions.   The measured optimum phase is 180°, in qualitative agreement with the theoretical value for zero growth rate, 240°.   Upon
stabilization, the plasma density increases, as shown by Fig. 4b.
Similar measurements and comparison with theory as a function of
microwave feedback gain are shown in Fig. 5.

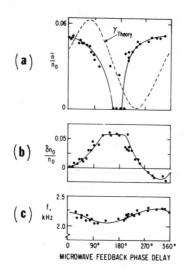

Fig. 4.   Instability amplitude (4a), change of plasma density at
the axis(4b), and instability frequency (4c), as a function of microwave feedback phase delay.   Dotted line in Fig. 4a is linear growth
rate calculated according to Eqs. (5) and (6).   Cesium plasma,
$T = 2800°$ K, $B = 4200$ G, $n_o = 5 \times 10^{10}$ cm^{-3}.

The radial distribution of feedback phase delay and feedback
gain is shown in Fig. 6 for an m= 2 mode.   (The plasma axis is
shifted to the left of the nominal axis due to the iron block.) Optimum
feedback phase is approximately 180° with deviations from this
value accounted for by the phase change of the instability in the
radial direction.   At both the near-side and the far-side maximum
of the instability amplitude (Fig. 1), where throughout the interaction
region the instability is nearly in phase, minimum feedback amplitude
is required.   These results also show that the <u>localized</u> energy source
used in the theoretical model is justified.

The feedback power calculated from linear equations for the
measured instability amplitude $(\tilde{n}/n_o \simeq 0.05)$, is $75\,\mu$W, in agreement
with the experimental value ($100\,\mu$W).

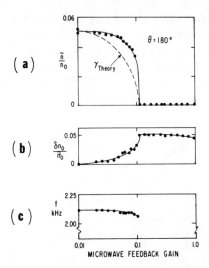

Fig. 5. Instability amplitude (5a), change of plasma density at the axis (4b), and instability frequency (4c) as a function of micro-wave feedback gain. Plasma parameters are the same as in Fig. 4.

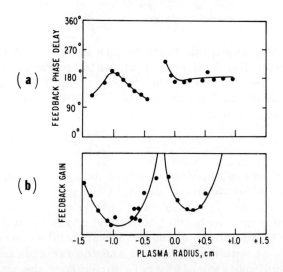

Fig. 6   Feedback gain (at optimum phase) and phase at different radial interaction region. Plasma parameters are the same as in Fig. 4.

V. DISCUSSION

This work demonstrates remote feedback suppression of collisional drift waves by modulated microwaves resonantly interacting
with plasmas at the upper hybrid frequency. The experimental
results are in qualitative agreement with linear-theory predictions
for a modulated energy source which, for measured amplitudes and
phases, reduces the instability growth rate by modulating the
electron temperature.

Recently, a similar experiment, using two microwave beams and
producing the low-frequency ion mode at the difference frequency,
or modulating one microwave beam at the ion-mode frequency, has
been reported (Wong, Baker and Booth, 1970).This experiment was
interpreted by a collisonless mode-coupling model for the following
reasons: a) the (high frequency) sum frequency is observed, b) the
electron temperature time-change was observed to be of the order of
1 msec, too slow to account for the rapid ion oscillations, and c) a
single side band excitation (produced by the nonlinear effects of one
single microwave) also excites the low frequency.

Low-frequency amplitude modulation of a microwave is equivalent
to two microwaves separated by the low frequency, i.e. all cases are
a three wave process (a), c)). The time change of the electron temperature is determined by the relaxation time and is much faster than
a drift wave period (b) (see also Stefanov (1970)). Although on the time
scale of the microwave frequency the particles are collisionless, the
three wave process includes the low (instability) frequency, on which
time scale the particles are strongly collisional and therefore max-
wellianized.

Theoretically, any complete non-linear treatment of the Vlasov
equation for electrons must also satisfy the moment equations. In the
presence of microwave heating the zeroth moment is

$$\frac{\partial n}{\partial t} + \nabla \cdot (n\underline{v}) = 0 \tag{7}$$

while the second moment is

$$\frac{3}{2} \frac{\partial}{\partial t} (nKT) + \nabla \cdot \underline{q} = \underline{E}_W \underline{j}_W \tag{8}$$

The momentum equation will be unaffected by microwaves because the
momentum imparted to the plasma by the absorption of microwaves is
negligible. As long as the electron-electron and electron-ion collision
frequencies are large compared to the wave frequency (as is true in our

experiments), the concepts of a scalar pressure and transport coefficients remain valid.   Hence the effect of microwaves appears
only in the energy equation.   Thus, experimental and theoretical
evidence indicates the validity of the temperature model.

ACKNOWLEDGMENT

    This work was performed under the auspices of the U. S.
Atomic Energy Commission, Contract No. AT(30-1)-1238.

3.3 EXPERIMENTS AND INTERPRETATION OF RESULTS ON THE FEEDBACK STABILIZATION OF VARIOUS PLASMA INSTABILITIES

B. E. Keen

UKAEA, Research Group, Culham Laboratory
Abingdon, Berkshire, England

ABSTRACT

Experimental results are reported on the feedback stabilization of a "drift-type" instability and on the remote sensing and feedback stabilization by amplitude modulation with electron-cyclotron resonance heating of ion-sound instability. For the "drift-type" instability, results are interpreted in terms of an equation of Van der Pol type; for the ion-sound instability in terms of a theory predicting the variation of instability amplitude and frequency due to the feedback modulation of electron temperature. Comparison between the theory and experiment show good agreement.

Part 1. Experiments on a "Drift-Type" Instability

I. INTRODUCTION

Recently a number of papers have shown feedback sup-pression of various "drift-type" plasma instabilities. Some success has been achieved in the interpretation of these results by the use of a feedback source term included in a linearized theory (Simonen, Chu, and Hendel, 1969). However, in order to allow for the positive feedback case where the oscillation is amplified, a nonlinear theory must be employed which limits the final instability level to a finite value. It has been shown that the Van der Pol type of equation (Van der Pol, 1934) gives a good description of certain nonlinear phenomena occurring in some plasma instabilities. These phenom-ena include mode locking and mode competition (Lashinsky, 1965). periodic pulling (Abrams, Yadlowsky, and Lashinsky, 1969); fre-quency entrainment or "synchronization" (Keen and Fletcher, 1969), and "asynchronous quenching" effects (Keen and Fletcher, 1970). This phenomenologocal approach is also adopted here, in which the Van der Pol equation is taken to describe the density oscillations in the plasma.

II. PHENOMENOLOGICAL THEORY

The equation in its simplest form without feedback is:

$$\frac{d^2 n_1}{dt^2} - (a - 3\beta n_1^2) \frac{dn_1}{dt} + \omega_o^2 n_1 = 0 \tag{1}$$

where n_1 is the density perturbation, ω_o the drift wave frequency, a the linear growth rate $(a/\omega_o \ll 1)$, and β a nonlinear saturation coefficient describing the final amplitude a_o of the oscillation:

$$a_o = (4a/3\beta)^{\frac{1}{2}} \tag{2}$$

Now consider a signal $gn_1(\tau)$ which is fed back, where g is an absolute gain in density perturbations and τ represents a delay time (here $\omega_o \tau = \phi$ the phase shift), then the equation is given by:

$$\frac{d^2 n_1}{dt^2} - (a - 3\beta n_1^2) \frac{dn_1}{dt} + \omega_o^2 n_1 + g \, \omega_o^2 n_1(\tau) = 0 \tag{3}$$

This equation is a simple example of a difference-differential equation. If a solution of the form $n_1 = a(t) \sin \omega t$ is assumed it can be shown that Eq. (3) can be brought into the form:

$$\frac{d^2 n_1}{dt^2} + \omega^2 n_1 = F[a(t), \omega] \cos \omega t + f[a(t), \omega] \sin \omega t \tag{4}$$
$$+ \text{ harmonics}$$

If the calculation is limited to the fundamental frequency ω the solution in the first approximation is:

$$a(t) = \frac{1}{2\omega} \int_o^t F[a(\varepsilon), \omega] d\varepsilon \quad ; \quad 0 = \frac{1}{2\omega} \int_o^t f[a(\varepsilon), \omega] d\varepsilon . \tag{5}$$

For the stationary state one has:

$$F(a, \omega) = 0 \quad , \quad f(a, \omega) = 0 \tag{6}$$

Hence, from Eqs. (4) and (6), the following conditions are obtained for the stationary state:

$$a_o^2 - a^2 = -g\left(\frac{\omega_o^2}{a\omega}\right) a_o^2 \sin \phi \quad , \tag{7}$$

$$\omega^2 = \omega_o^2(1 + g \cos \phi) . \tag{8}$$

Equation (7) shows that as the gain g is increased, the amplitude a will increase or decrease according to the sign of $\sin \phi$. Optimum suppression is achieved with

$$\sin \phi = -1 \quad (\text{i.e. } \phi = -90^{\circ} \text{ or } +270^{\circ}) \qquad ; \quad (9)$$

then suppression occurs when $g = \alpha/\omega_o$ (since $\omega = \omega_o$ at $\phi = -90^{\circ}$).

III. EXPERIMENTS

The apparatus was the same as that employed previously (Keen and Aldridge, 1969). The plasma was a hollow cathode arc discharge in argon, with electron temperature ~ 5.0 eV, peak density ~ 10^{13} cm^{-3}, and inverse density-gradient scale length - $1/n_o$ ($\partial n_o/\partial r$) = 0.70 ± 0.05 cm^{-1} in an axial magnetic field of 1 kG. The instability was predominantly an m = +1 instability with long axial wavelength λ (λ > 200 cm); its frequency was ~ 7.0 kHz. It was identified as a collisional type drift wave.

Density perturbations were detected on an ion-biased probe, and this signal was fed back via a wideband amplifier, phase shifter (variable over 450°), and power amplifier, to a plate in the plasma. This plate was in the same axial plane as the detecting probe and could be moved radially across the plasma. Minimum gain was required for suppression when the plate was situated at the radius (r ≈ 1.1 cm) corresponding to maximum instability amplitude. However, experiments were usually performed with the plate at r = 1.6 - 2.0 cm so that it caused less disturbance to the plasma. Feedback effects were observed on a further ion biased probe which could be moved axially and radially, and the output was displayed on a spectrum analyzer.

IV. RESULTS

The phase angle ϕ in the feedback loop was varied until a minimum was obtained in the instability level a, and then the ampli- tude a was measured as a function of the gain G in the wideband amplifier. The variation of $(a/a_o)^2$ plotted against increasing gain G, is shown in Fig. 1, for the suppressor plate set at two different radii r = 1.6 ± 0.1 cm, and r = 2.0 ± 0.1 cm. It is seen that a good linear relationship is obeyed, as predicted by Eq. (7).

Further experiments were performed with plate at radius r ≈ 1.8 ± 0.1 cm. The gain G was left set at its value for suppres- sion (G = 25.2) and the phase angle ϕ varied through 360°, and the

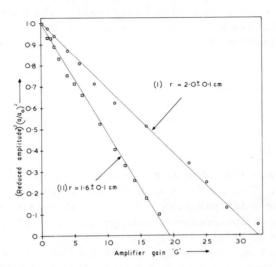

Fig. 1.   The square of the reduced amplitude $(a/a_o)^2$ plotted
versus amplifier gain G for the conditions when the suppressor
plate is set at a radius (r) such that (1) r = 2.0 ± 0.1 cm and (2)
r = 1.6 ± 0.1 cm.

relative amplitude (a/a_o) measured.   This is shown plotted in
Fig. 2(a) as $(a/a_o)^2$ versus phase angle ϕ.   Optimum suppression
is achieved when $\phi = -90^\circ$, or $+270^\circ$ as predicted by Eq.   (9).
Figure 2(b) shows $(a/a_o)^2$ plotted versus $\sin\phi$, and a good linear
relationship is seen to be obeyed as predicted.   Other gain values of
G = 12.6 and 7.9 are shown plotted in both Figs. 2(a) and 2(b).

The absolute gain g of the system was calibrated with the aid
of Eq. (8).   For optimum suppression (G = 25.2) theory gives
$g = \alpha/\omega_o \ll 1$, and, also if $(\omega - \omega_o) = \Delta\omega$, Eq. (8) shows that
$2\Delta\omega/\omega_0 = g\cos\phi$.   The measured frequency shift $\Delta\omega$ plotted as a
function of $\cos\phi$ is shown in Fig. 3, and the slope of this line is
proportional to $g = \alpha/\omega_o (= 0.12 \pm 0.02)$.   Knowing the absolute gain
g, one can calculate the theoretical variation of $(a/a_o)^2$ as function
of phase angle ϕ for each gain value, using Eq. (7).   This variation
is shown as the continuous lines in Fig. 2(a).   Finally, a further
check on the theory was made by measuring directly the linear
growth rate α and comparing it with the value obtained indirectly.
This was effected by using a tone-burst generator in the return loop,
which gated the feedback signal at periodic intervals.   The resulting
instability signal was analyzed, and gave an average value for the

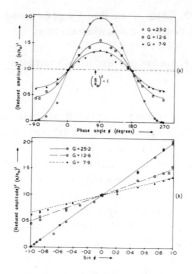

Fig. 2. The square of the reduced amplitude $(a/a_o)^2$ plotted against (a) phase angle ϕ and (b) $\sin\phi$.

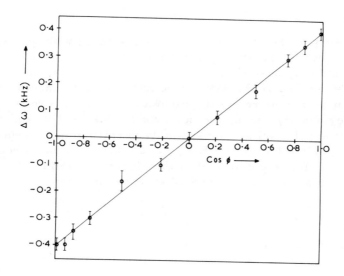

Fig. 3. The change in the instability frequency $\Delta\omega$(kHz) plotted versus the cosine of the phase angle $(\cos\phi)$.

growth rate α of $(0.14 \pm 0.03)\omega_o$, in good agreement with the value obtained above from theoretical considerations.

Part 2. Experiments on an Ion-Sound Instability

I. INTRODUCTION

Most of the previous work on feedback stabilization of various plasma instabilities has been performed with feedback suppressor elements in contact with the plasma. However, if feedback control of instabilities is to be a feasible method in any thermonuclear plasma, remote sensing and suppressor elements are essential. Recently, experiments have been performed (Hendel, Chu, Perkins, and Simonen, 1970; Wong, Baker, and Booth, 1970) in which a remote modulated microwave source, irradiating the plasma at the upper hybrid frequency, has been used as the feedback element. Here, experiments are reported in which an m = 0 ion-sound instability has been suppressed by using a remote source of energy at the electron-cyclotron resonance (E.C.R.) frequency, amplitude modulated at the instability frequency. The effect of the energy at E.C.R. is to cause a local increase in the temperature, ΔT, which in this case is small compared with the steady state electron temperature, T_e.

II. THEORY

The stability of the plasma has been considered with the two-fluid model, in which the axial magnetic field, B_o, is taken in the z direction, and only spatial variations of the form $\exp(ik_z z)$ are considered, where k_z is the axial wave number of the instability. The density n is considered of the form $n = n_o + n_1$, where n_o is the zero order density and n_1 the perturbed value, and V_1 and v_1 are taken as the potential and ion-velocity perturbations, respectively. If the main effect of the E.C.R. is taken to be local heating of the electrons, the z component of the electron equation of motion reduces in the low frequency approximation to the form:

$$eV_1 = T_e (n_1/n_o) + f\Delta T \langle n_1 (\tau)/n_o \rangle \qquad (10)$$

where f is the fractional modulation produced by the feedback signal and $n_1(\tau)$ represents the density perturbation delayed by a time τ.

The ion equation of motion is:

$$\frac{d\bar{v}_i}{dt} = -\frac{e}{M_i} \bar{\nabla} V_1 - \bar{v}_1 \nu \quad \frac{e}{M_i} [\bar{v}_1 x \bar{B}_o] \qquad (11)$$

where $1/\nu$ is the ion-neutral collision time, and the equation of continuity is given by:

$$\frac{\partial n}{\partial t} + \bar{\nabla} \cdot (n\bar{v}_1) = S_i(n_1) \tag{12}$$

S_i is a source term due to ionization, etc., caused by large amplitude oscillations present in the plasma. This source term is taken to be of the form:

$$S_i = \alpha n_1 - \gamma n_1^2 - \beta n_1^3 \tag{13}$$

where $\beta n_1^2 \ll \gamma n_1 \ll \alpha \ll \omega_0 = k_z C_s$, and $C_s = (T_e/M_i)^{1/2}$ is the ion-sound velocity. After eliminating v_1 between (11) and (12) and substituting for S_i for (13), in the approximation that ν is small, the equation for n_1 reduces to

$$\frac{d^2 n_1}{dt^2} - \frac{dn_1}{dt}[\alpha - 2\gamma n_1 - 3\beta n_1^2] + \omega_0^2 n_1 + \omega_0^2 f(\frac{\Delta T}{T_e}) n_i(\tau) = 0 \tag{14}$$

which is the Van der Pol equation including a feedback term similar to Eq. (3). This may be solved in a similar way to give:

$$(a_0^2 - a^2) = -f(\frac{\Delta T}{T_e})(\frac{\omega_0^2}{\alpha\omega}) a_0^2 \sin \phi \tag{15}$$

and

$$\omega^2 = \omega_0^2 \left[1 + f(\frac{\Delta T}{T_e}) \cos \phi \right] \tag{16}$$

where a_0 and a, are the instability amplitudes, and ω_0 and ω the frequencies, with and without feedback, respectively. Again, Eq. (15) shows that as the fractional modulation of the E.C.R. frequency (proportional to the gain in the feedback loop) is increased the amplitude a will increase or decrease according to the sign of $\sin \phi$. Optimum suppression is achieved for

$$\sin \phi = -1, \text{ i.e. } \phi = -90^\circ \text{ or } + 270^\circ . \tag{17}$$

Then

$$f(\Delta T/T_e) = a/\omega_0 \tag{18}$$

where α is in fact the linear growth rate.

III. EXPERIMENTS

The apparatus used was similar to that employed previously
(Keen and Fletcher, 1969; Keen and Fletcher, 1970). The plasma
was the positive column of a neon arc discharge, and had a peak
density $\sim 3 \times 10^{11}$ cm^{-3}, an electron temperature T_e = 5.4 eV, and
was contained in a magnetic field of $\sim$ 180 G. A remote photo-diode
was employed to sense the instability, which was found to have
mainly a single frequency $\sim$ 7.5 kHz independent of magnetic field,
and an m = 0 azimuthal mode number. It was identified as an ion-
sound instability.

Power at the E.C.R. frequency was applied to the plasma by a
Lisitano-type resonant structure (Lisitano, 1966) around the glass
tube, from a U.H.F. oscillator and power amplifier. The temperature change ΔT produced was linearly proportional to the power
input to the cavity in the range used in these experiments. A signal
proportional to the density perturbations was fed from the photo-
diode via a phase shifter, and a L.F. amplifier, to amplitude
modulate the U.H.F. oscillator at the E.C.R. frequency. The
effect of the feedback was observed by displaying the output from a
separate probe on a spectrum analyzer.

IV. RESULTS

With the E.C.R. power on, the phase angle ϕ in the feedback
loop was varied until minimum instability amplitude was achieved.
Then, the gain in L.F. amplifier was varied until the fractional
modulation was $\sim$ 1.0 and the power level at the E.C.R. frequency
was set so that suppression just occurred. At this value, with only
the E.C.R. frequency on, the measured change in electron temperature was $\Delta T = 0.6 \pm 0.1$ eV. At this phase angle and E.C.R. power
level, the amplitude a was measured as a function of the gain in
the L.F. circuit (αf). Figure 4 shows $(a/a_o)^2$ versus fractional
modulation f, and it is seen that a good linear relationship is
obtained as predicted by Eq. (15). In Fig. 4 photograph (a) shows
the instability when f = 0 and photograph (b) shows the effect when
it is suppressed with f = 1.0.

At the same phase angle, and with the fractional modulation
maintained at 1.0, the instability amplitude a was measured as a
function of the input power at the E.C.R. frequency $\alpha \Delta T$. This is
shown plotted in Fig. 5 as $(a/a_o)^2$ versus E.C.R. power $(\alpha \Delta T)$, and
shows a linear relationship as predicted by Eq. (15). Figure 6,
shows the square of the reduced amplitude $(a/a)^2$ plotted against
phase angle ϕ in the feedback loop, and it is seen that the suppression occurs at a phase change of $-105^o \pm 20^o$ or $255^o \pm 20^o$ as

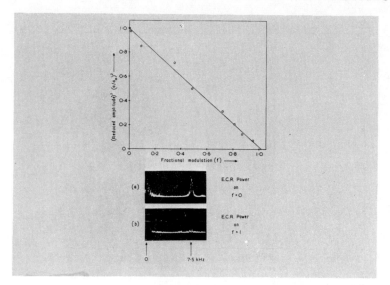

Fig. 4. The square of the reduced amplitude $(a/a_o)^2$ versus the fractional amplitude modulation, f. Photographs show the spectrum analysis of the instability for (a) f = 0, and (b) f = 1.0.

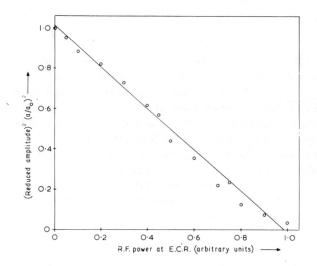

Fig. 5. The square of the reduced amplitude $(a/a_o)^2$ plotted versus the input power at the E.C.R. frequency $(\alpha \Delta T)$.

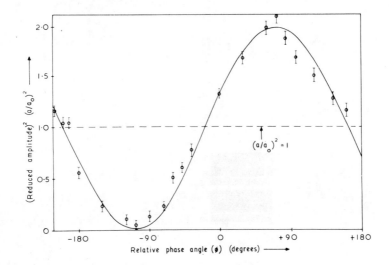

Fig. 6. The square of the reduced amplitude $(a/a_o)^2$ plotted versus phase angle ϕ, in the feedback loop.

compared with the predicted value [Eq. (17)] of -90° or 270°. The full line in Fig. 6 is the theoretical predicted variation of Eq. (15) shifted so that its minimum is at -105° instead of -90°. Here again, good agreement is obtained.

From Eq. (15), it is possible to obtain a value for α, the linear growth rate. At the position for optimum suppression (a = 0) and $\alpha/\omega_o = f(\Delta T/T_e) = 0.11 \pm 0.02$. This value was checked by directly measuring the growth and decay rate of the instability amplitude. This was effected by suppressing the instability by the method of "asynchronous quenching" (Keen and Fletcher, 1970). By gating the asynchronous frequency the instability was turned on and off at periodic intervals. Analysis of the resulting signal gave $\alpha/\omega_o = 0.12 \pm 0.03$, in good agreement with the former value. A further check was afforded by the use of Eq. (16). If the frequency shift $\Delta\omega = (\omega - \omega_o)$ is measured as a function of phase angle, the maximum frequency shift $\Delta\omega_m$ is given by $2\Delta\omega_m/\omega_o = f(\Delta T/T_e) = \alpha/\omega_o$. The measured maximum change in frequency was $\Delta\omega_m = 0.3 \pm 0.1$ kHz, and thus the value $\alpha/\omega_o = 0.08 \pm 0.03$ resulted which is not in such good agreement.

CONCLUSIONS

Adopting a nonlinear equation of the Van de Pol type to explain the instability saturation conditions in a plasma, relationships can be obtained between the amplitude a and frequency shift $(\Delta\omega)$ as a function of gain and phase shift ϕ in the feedback loop. The resulting measurements show the predicted variations and a consistent value for the growth rate α is obtained within experimental error.

3.4 FEEDBACK SUPPRESSION OF COLLISIONLESS, MULTI-MODE DRIFT-CENTRIFUGAL WAVES IN A MIRROR-CONFINED PLASMA[*]

N. E. Lindgren and C. K. Birdsall
Electronics Research Laboratory, University of California,
Berkeley, California 94720

ABSTRACT

Simultaneous suppression of several modes of the drift-centrifugal wave in a collisionless plasma has been achieved by using as many independent feedback systems, each one tuned to a different mode. This is an improvement over broadband systems in which suppression of one mode has generally led to enhancement of other modes.

I. INTRODUCTION

Feedback stabilization has recently been reported for drift waves (Simonen, Chu, and Hendel, 1969) and drift-like instabilities (Parker and Thomassen, 1969; Keen and Aldridge, 1970) in collision dominated plasmas, where only one mode existed or was isolated by careful adjustment of the plasma parameters. For collisionless plasmas, we and others (Politzer, 1969; Buchel'nikova, 1964; Little and Barrett, 1967) typically find that many azimuthally varying modes are present simultaneously; isolation of one mode is difficult or impossible. Thus simultaneous suppression of many modes is clearly the more general problem in plasmas. The significant result of this work is that by using filtered feedback loops, one to a mode, the modes can be independently and simultaneously suppressed.

II. EXPERIMENT

The apparatus on which the experiments were performed is the Berkeley Plasma Instability Experiment (P.I.E.), in which a potassium plasma is produced by contact ionization in a 50 cm-long mirror magnetic field of mirror ratio 2.5. The density used was about 10^8 cm^{-3}, ion and electron temperatures were about 0.2 eV., and the magnetic fields at the midplane ranged from 0.5 to 1.0 kG. At this temperature and density the unstable waves are considered collisionless since ion collision frequencies are at least one order of magnitude

---

[*]Work supported by U.S. Atomic Energy Commission, Contract AT(11-1)-34, P.A. 128.

lower than the observed frequencies. A dc radial electric field
causes the plasma to rotate at speeds comparable with and larger
than (up to five times) the diamagnetic speed.

The waves to be suppressed by feedback have many of the properties of the gravitation-like centrifugal instability (Rosenbluth,
Rostoker and Krall, 1962; Chen, 1966; Chu, Coppi, Hendel, and
Perkins, 1969) resulting from the $\underline{E}_r \times \underline{B}$ plasma rotation. The
azimuthal propagation in the (rotating) plasma frame is in the diamag-
netic current direction; $\delta\phi$ leads δn by about 100° in the propagation
direction at the radius of maximum wave amplitude; the waves have
maximum amplitude near the radius of maximum $(-1/n)(dn/dr)$; the
instability vanishes if the plasma rotation speed is made sufficiently
small by applying a potential between the cathode-ionizer and the
vacuum chamber so as to reduce the radial electric field. The wave
spectrum analysis shows up to eight discrete frequencies ($\omega < \omega_{ci}$)
corresponding to azimuthal modes, $m = 1$ to 8. The lowest four modes
were identified by phase comparison on several azimuthally spaced
probes. All azimuthal modes have maximum amplitude at the same
radius and $\delta n/n_0 \approx \delta\phi/KT \approx 0.05$. $k_\parallel \ll k_\perp$; typically $k_\parallel/k_\perp \approx 0.05$
for the $m = 2$ mode. These waves are observed in a steady, presum-
ably saturated state.

III. RESULTS

Feedback suppression was tried with several loops, Fig. 1.

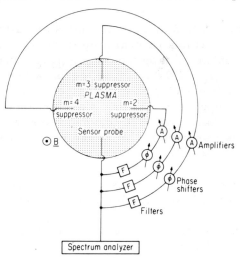

Fig. 1. Schematic of feedback arrangement. Each feedback
system consists of a filter, phase shifter, amplifier, and a suppressor
probe. Each system is tuned to a different mode.

The feedback probes were located at the position of maximum instability amplitude. Attempts at stabilization when two or more modes
are present, using amplifier(s) with bandwidth(s) sufficient to over-
lap the frequencies of the several modes, have tended to be unsucces-
sful. We and others (Simonen, Chu and Hendel, 1969; Parker and
Thomassen, 1969) have found that as one mode is suppressed, one or
more of the other modes are usually enhanced, as illustrated in
Fig. 2a; with three modes (m = 1, 2, 3) present, the feedback phase
was varied through 360° for the dominant mode (m = 2, f = 15 kHz),
with the alternate suppression and enhancement of each mode.

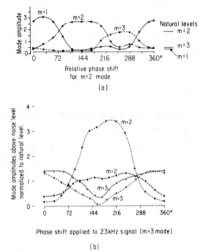

Fig. 2. (a) m = 1, 2, 3 azimuthal modes are alternately suppressed and built up (to 7 times natural amplitude for the m = 1 mode)
when broadband feedback is applied. $B_{midplane}$ = 0.67 kG, n =
1×10^8 cm^{-3}, $T_{cathode}$ = 2100° K. The feedback loop gain was about
2000, which gave a feedback probe voltage of 2 V prior to suppression.
(b) Comparison of suppression of m = 3 and enhancement of m = 2
when broadband feedback was used (dashed lines) and when m = 3
pass-band feedback was used (solid lines). Broadband feedback
greatly enhanced m = 2.

In such experiments usually there is little decrease in the total wave
energy, so that over-all stabilization is not achieved.

By limiting the bandwidth of the signal to be amplified in the feedback loop, using the filters shown in Fig. 1, so that each loop tends to
operate on only one mode, we have been able to achieve suppression
without enhancement.

When only one feedback system was used, representative results are shown in Fig. 2b, with and without filtering. Without filtering, the m = 3 mode can be reduced to a few percent of its natural (no feedback) level where, however, the m = 2 mode is increased by a factor of more than 3 so that the total wave energy is up by about a factor of 10. With filtering, it is seen that the m = 3 mode can be reduced to about 30 percent of its natural level, with only slight enhancement of the m = 2 mode (about 10 percent), with total wave energy down about 30 percent. In other tests, suppression with no enhancement has been obtained.

When three feedback systems are used, representative results are given in Fig. 3.

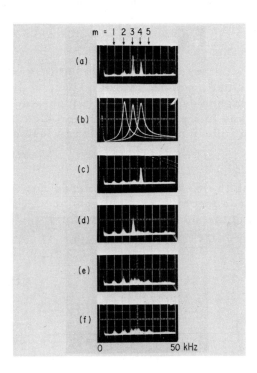

Fig. 3. (a) Spectrum of natural signal. (b) Amplitude charac-teristic of three filters tuned to pass the m = 2, 3, 4 frequencies. (c) m = 3 feedback system operated alone suppressed m = 3 without significantly enhancing any other modes. (d) m = 4 feedback system operated alone suppressed m = 4. (e) m = 3 and 4 systems operated simultaneously. (f) m = 2, 3, 4 systems operated simultaneously suppressed all the low frequency modes to within a few times the noise level.

The natural mode spectrum is shown in (a), dominated by modes 3 (22 kHz.) and 4 (27 kHz.). The three filters were tuned as shown in (b) to peak at the m = 2, 3, 4 mode frequencies; the filter Q's were made to be roughly 7 such that the voltage loop gains at neighboring modes were down by at least 5. Using large amplifier gain only in the m = 3 loop this mode was suppressed, as in (c); with large gain only in the m = 4 loop, the result is similar, as in (d). Little or no enhancement of other modes is observed. With large gains in the m = 3 and m = 4 loops, these modes are both suppressed, as in (e), with some enhancement of the m = 2, 5 modes. By adding large gain to the m = 2 loop, this mode was reduced from its value in (e), as shown in (f). This sequence shows that the use of independent narrowband feedback loops has resulted in reducing one to three low frequency modes to within a few times the local noise value, with little enhancement of the other modes.

IV. CONCLUSIONS

Because the parameters in this experiment (e. g. $\omega_{pi}^2/\omega_{ci}^2 \gg 1$, a_i (-1/n)(dn/dr) not small, non-Maxwellian $f_i(\underline{v})$, collisionless ions, etc.) scale to much higher density, higher temperature plasmas, encouragement is given to the solution of the problem of low frequency ($\omega \lesssim \omega_{ci}$) multiple-mode stabilization in fusion plasmas.

ACKNOWLEDGEMENT

The authors wish to thank Dr. H. W. Hendel for the original encouragement in these feedback experiments and M. J. Bales for designing and building the necessary electronics.

3.5 FEEDBACK APPLICATIONS TO BASIC PLASMA INSTABILITY EXPERIMENTS

Thomas C. Simonen

Princeton University, Princeton, New Jersey 08540

ABSTRACT

Feedback has been applied, as a diagnostic technique, to measure properties of low-frequency instabilities. Feedback-stabilized plasmas provide equilibria in which the temporal growth of instabilities to non-linear saturation levels and the effect of instability on confinement can be measured by switching feedback off. Enhancement of instability amplitudes by positive feedback above the decay-instability threshold value, can be employed to destabilize parametrically excited decay modes. Experiments with theories illustrating such feedback applications to the collisional drift instability are described.

I. INTRODUCTION

Suppression of instabilities in thermonuclear plasmas has motivated many experiments (Arsenin, Zhil'tsov, and Chuyanov, 1969; Arsenin, Zhil'tsov, Likhtenstein, and Chuyanov, 1968; Chuyanov, 1969; Parker and Thomassen, 1969; Keen and Aldridge, 1969; Simonen, Chu, and Hendel, 1969; Chuyanov, Murphy, Sweetman, and Thompson, 1969; Hendel, Chu, Perkins, and Simonen, 1970; Wong, Baker, and Booth, 1970; Lindgren and Birdsall, 1970) in plasma feedback control. A second important application of feedback is as a diagnostic technique to measure properties of low-frequency plasma instabilities. This paper describes such diagnostic feedback applications. To maintain a degree of uniformity, most of the results presented are from Princeton Q-1 machine collisional drift wave experiments, although the methods can be, and in many instances have been, applied to other instabilities in other devices. Characteristics of the drift instability in Q-devices are described in earlier references (Hendel, Chu, and Politzer, 1968; Rowberg and Wong, 1970). In the following sections we describe feedback applications to measurements of (a) instability growth rates and amplitude saturation coefficients, (b) effects of instability on plasma confinement, and (c) parametrically excited decay-instabilities.

II. INSTABILITY GROWTH RATES

Feedback has been employed to measure linear growth rates and nonlinear saturation coefficients of instabilities in the single-mode regime (Arsenin, Zhil'tsov, and Chuyanov, 1969; Chuyanov, 1969; Wong and Hai, 1969; Chu, Hendel, Schlitt, Simonen, and Stix, 1969 ; Keen, 1970). The time development of instability amplitude, $a(t) = A(t) \cos \omega_0 t$, can be represented by the amplitude equation (Landau and Lifshitz, 1959)

$$dA^2/dt = (2\gamma - \alpha A^2 - \sum_i \beta_i A^{2i+2}) A^2 \qquad (1)$$

where γ is the linear growth rate and α, β_i are nonlinear coefficients. For $\beta_i = 0$, Eq. (1) describes the Van der Pol (1934) equation; its amplitude development is:

$$A^2(t) = \frac{2\gamma}{\alpha} \left\{ 1 + e^{-2\gamma t} [2\gamma/\alpha A^2(0) - 1] \right\}^{-1} . \qquad (2)$$

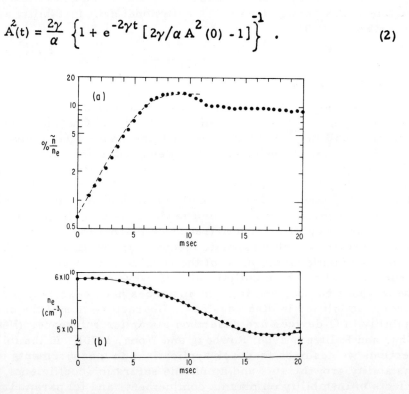

Fig. 1. (a) Drift instability amplitude and (b) equilibrium density vs time after switching feedback off in a potassium plasma with B = 2240 Gauss = 1.03 B_c. Dashed curve is Eq. (2) with $\gamma = 0.5$ msec^{-1} and $\alpha = 40$ msec^{-1}.

An m = 2 density wave signal as a function of time, is shown in Fig. 1 after the feedback stabilization is switched off. The instability grows exponentially until $A(t) > \frac{1}{2} A (\infty)$, and then saturates according to Eq. (2). As the instability grows, wave induced losses reduce the equilibrium density on a slow time scale allowing the instability to overshoot its steady-state amplitude. Collisional drift instability growth rates versus magnetic field strength (Chu, Hendel, Schlitt, Simonen, and Stix, 1969) measured by this feedback technique, together with theoretical values (Hendel, Chu, and Politzer, 1968) are shown in Fig. 2.

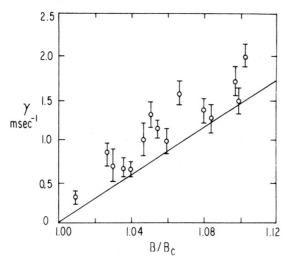

Fig. 2. Measured m = 2 linear growth rate vs magnetic field normalized to onset value B_c = 2100 G. Theoretical line normalized to theoretical B_c ≈ 1450 Gauss.

Nonlinear saturation coefficients can be determined by plotting $A^{-2} \, dA^2/dt$ versus A^2 and finding γ, α, β_i by curve fitting to Eq. (1). The data of Fig. 1, plotted in this manner in Fig. 3, yields the linear growth rate, $\gamma = 0.5 \, \text{msec}^{-1}$, and the first nonlinear saturation coefficient, = 40 msec^{-1}, in qualitative agreement with theoretical values.

Feedback effects $F \, (\omega, k)$ may be incorporated into the linear dispersion relation $D(\omega_o, k_o) = 0$ in the form (Parker and Thomassen, 1969; Keen and Aldridge, 1969; Furth and Rutherford, 1969; Chen and Furth, 1969; Taylor and Lashmore-Davies, 1970)

$$D(\omega, k) = F \, (\omega, k) = \sigma e^{i\Theta} f(\omega, k) \qquad (3)$$

from which one finds that feedback perturbs the instability eigenfrequency ω_0 (to first order in σ)

$$\omega = \omega_0 - \sigma e^{i\theta} f(\omega_0, k_0) \left(\partial D/\partial \omega \big|_{\omega_0, k_0}\right)^{-1} \tag{4}$$

as illustrated in Fig. 4.   Optimum stabilization occurs when $\theta = \theta_{opt} = \pi/2 - \arg(f \cdot dD/\partial \omega) \approx 240°$ as shown in Figs. 4 and 5.

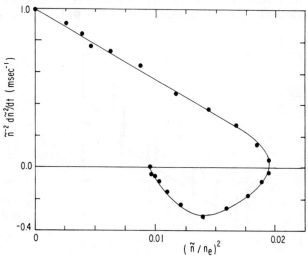

Fig. 3.   Nonlinear growth rate vs instability amplitude.

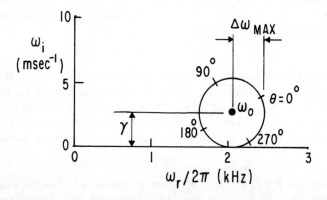

Fig. 4. Feedback variation of instability eigenfrequency in complex ω-plane. Circular mapping is for $\sigma = 5.75$ msec^{-1} and $0° < 360°$ in a potassium plasma, with $n_e = 10^{11}$ cm^{-3}, and $B = 1.17 B_c$.

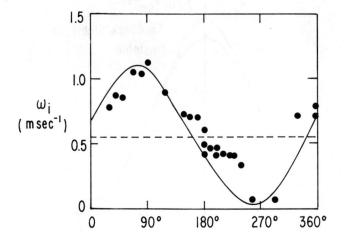

Fig. 5.   Measured feedback-controlled drift-instability growth-rate vs feedback phase.   Theoretical curve, with minimum feedback stabilizing gain, calculated for experimental parameters.

The maximum frequency shift $\Delta\omega_{max}$, associated with minimum stabilization gain at $\theta = \theta_{opt} \pm \pi/2$, may be employed, as illustrated in Fig. 4, to indirectly determine the instability growth rate, $\gamma = \Delta\omega_{max}$.   Such growth-rate measurement has been done by Keen (1970).   The measured frequency shift $\Delta\omega_{max}$ is in agreement[†] with the directly measured growth rate $\gamma = 2.6\,\text{msec}^{-1}$.

III.   INSTABILITY INDUCED LOSS

Instability suppression (e.g. by variation of magnetic field strength (Hendel, Chu, and Politzer, 1968) or by adding shear (Mosher and Chen, 1970) permits wave induced loss to be separated from other losses.   Thus, feedback may be employed to measure instability induced loss.   Increase in plasma density with feedback stabilization of the collisional drift wave is shown in Fig. 6.

Positive feedback may be employed to destabilize waves which are stable for given plasma parameters, as shown in Fig. 7, in order to assess the effects of potentially unstable modes on plasma confinement.

[†]The paper by Keen (1970) contains an error; nevertheless the data show that the two growth-rate measurements are in agreement. Private communication, B.E. Keen.

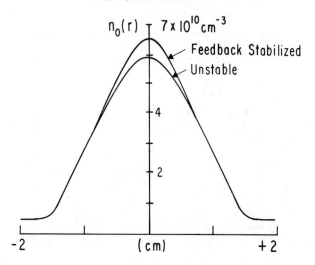

Fig. 6.   Radial distribution of plasma equilibrium density with
and without feedback stabilization.

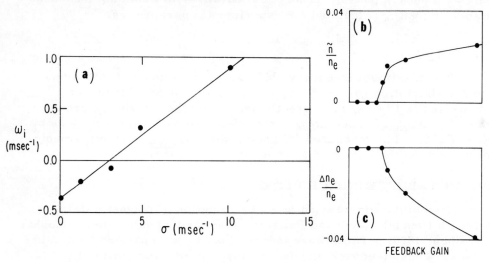

Fig. 7.   (a) Collisional drift wave growth rate,   (b) amplitude and (c) change of central density of naturally stable plasma
column vs positive feedback gain.

Fig. 8 shows equilibrium density decrease in the center of the
plasma column versus the square of the feedback controlled drift
wave amplitude together with the equation (Chu, Hendel and Politzer,
1967)

$$\frac{\Delta n_e}{n_e} = \frac{m}{\alpha_L} \frac{kT}{e_B} \frac{1}{r^2} \frac{\tilde{n}}{n_e} \cdot \frac{e\tilde{\phi}}{kT} \sin \psi. \tag{5}$$

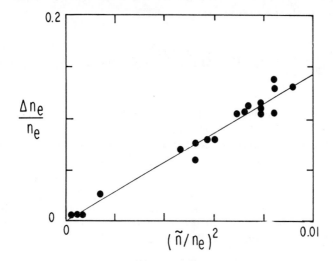

Fig. 8.   Relative density decrease versus square of normalized m = 2 instability amplitude.   Curve is Eq. (5) for α_L = 380 sec^{-1}, T = 2700°K, B = 2000 Gauss and ψ = 9°. ψ_{meas} = 10°- 20°.

For this weakly unstable plasma, γ/ω = 0.024, neither the $\tilde{n}$ - $\tilde{\phi}$ phase difference ψ nor the ratio $\tilde{n}/\tilde{\phi}$ are affected by feedback (Simonen, Chu, and Hendel, 1969; Furth and Rutherford, 1969).

$$\frac{\tilde{n}}{n_e} = \frac{1/t_\| - i\omega_e}{1/t_\| - i\omega + \sigma} \frac{e\tilde{\phi}}{kT} \tag{6}$$

Therefore, the density decrease follows $\tilde{n}^2$.

IV.   DECAY INSTABILITIES

Two linearly stable waves (ω_1, k_1) and (ω_2, k_2) can be parametrically excited (Sagdeev and Galeev, 1969; Wong and Hai, 1969) by a higher-frequency, large-amplitude pump wave (ω_o, k_o) if all three waves satisfy (a) the linear dispersion relation, (b) coupling conditions $\omega_o = \omega_1 + \omega_2$ and $k_o = k_1 + k_2$ and (c) if the pump wave amplitude exceeds a threshold amplitude, $\phi_o > \phi_m$.   Decay processes satisfying

all of the conditions have been observed for collisional drift waves
in the stable regime (Wong and Hai, 1969) and electron cyclotron
harmonic waves (Chang and Porkolab, 1970). For the degenerate drift
wave case, $\omega_1 = \omega_2$, $k_1 = k_2$, the decay wave nonlinear growth rate can
be expressed, with feedback, as

$$\gamma_{NL} = \gamma + (\gamma^2 \phi_0^2/\phi_m^2 - (\omega_0/2 - \Omega)^2)^{1/2} + G \cos(\theta - \theta_0) - \tfrac{1}{2}\alpha\phi_1^2. \quad (7)$$

Here γ is the linear growth (or damping) rate, ϕ_0 and ϕ_1, the pump
and decay mode amplitudes, Ω the decay mode eigenfrequency, G and
θ the feedback gain and phase, and α the decay mode nonlinear saturation coefficient. It is evident from Eq. (7) that the decay mode may
be feedback controlled not only by the feedback parameters (G,θ) but
also by feedback variation of the pump-wave amplitude ϕ_0 and frequency ω_0. Using high and low pass filters, one can thus selectively
apply feedback to : (a) the pump mode or (b) the decay mode, to probe
the parametric three wave coupling process.

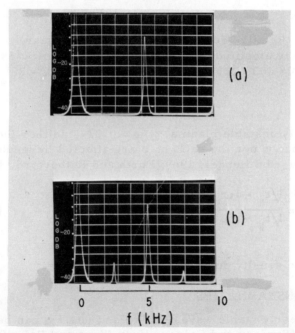

Fig. 9. (a) Collisional drift instability spectra without feedback
and (b) spectra, showing 2.5 kHz decay mode, with positive feedback
applied to 5 kHz pump mode.

Figure 9 shows the destabilization of the m = 1 degenerate decay-instability by positive feedback enhancement of the m = 2 collisional drift-instability amplitude above its decay threshold.

Feedback may be employed to measure degenerate decay-instability threshold amplitudes (Chu, Hendel, and Simonen, 1970) by exciting the plasma with a signal at the instability subharmonic $1/2\,\omega_0$. The plasma response ϕ to such a feedback excitation, with strength μ, may be calculated from the theory of Stix (1969)

$$\phi = \frac{\mu}{A}\ \frac{1 - (\phi_0/\phi_m)\ \exp\ (-i\theta)}{1 - \phi_0^2/\phi_m^2} \tag{8}$$

Here θ is the excitation phase referenced to the pump wave and A is the excited mode linear dielectric function. By varying the excitation phase, as illustrated in Fig. 10, parametric pumping and suppression is measured from which the threshold amplitude is determined, $\phi_m = \phi_0(1 + \phi_{min}/\phi_{max})\ /\ (1 - \phi_{min}/\phi_{max})$.

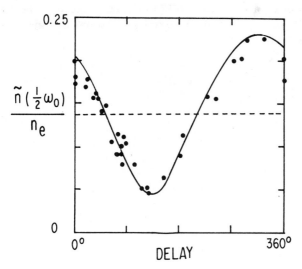

Fig. 10. Amplitude of excited degenerate m = 1 mode $(\omega=1/2\omega_0)$ vs phase relative to m = 2 pump mode $(\omega = \omega_0)$. Curve is Eq. (8) calculated for $\phi_m = 1.3\ \phi_0 = 1.3\tilde{n}/n_e = 0.09$ and arg $(\phi_0/\phi_m) = -135°$.

V. SUMMARY

A variety of experiments, employing feedback to measure growth rates, wave induced losses, and parametric decay instabilities associated with collisional drift waves, has been described. Many of these techniques have also been applied to study other instabilities including the flute (Arsenin, Zhil'tsoz, and Chuyanov, 1969), ion cyclotron (Arsenin, Zhil'tsov, Likhtenstein, and Chuyanov, 1968), ionization (Garscadden and Bletzinger, 1969), Kelvin-Helmholtz (Chu, Hendel, Jassby, and Simonen, 1970; Müller, Corbin, and Palmer, 1970), radial-electric-field driven drift instability (Parker and Thomassen, 1969; Keen and Aldridge, 1969), and current driven instabilities in electron-hole plasmas (Ancker-Johnson, 1970). These experiments demonatrate that feedback provides a new and powerful technique to probe plasma instabilities.

ACKNOWLEDGMENTS

It is a pleasure to thank my colleagues Drs. H. W. Hendel and T. K. Chu, with whom this work was originated and excuted. The help of P. Blaney, L. Gereg and L. Schlitt with the experiment and H. Fishman with the numerical calculations is also acknowledged.

This work was performed under the auspices of the U.S. Atomic Energy Commission Contract No. AT(30-1)-1238; also, use was made of computer facilities supported in part by National Science Foundation Grant No. NSF-GP-579.

3.6 CHARACTERISTICS OF A MODIFIED FEEDBACK CONTROL METHOD APPLIED TO EDGE OSCILLATIONS IN A Q-MACHINE

G. L. J. Müller[*], J. C. Corbin, and R. S. Palmer
Aerospace Research Laboratories, WPAFB, Ohio, 45433

ABSTRACT

Experimental results on the feedback control of an "edgetype" oscillation in a Q-machine, using a fixed-amplitude drive signal phase- and frequency-locked to the plasma oscillation, are reported. Results are compared with stationary solutions of a theoretical model described by a second order linear differential equation. In connection with this theory, growth rates and a-c coupling between suppressor probe and plasma are measured and compared with calculated values. The method provides a means of measuring radial plasma losses caused by this instability.

I. INTRODUCTION

A number of studies have been made using feedback control to stabilize and suppress low frequency instabilities in various plasmas (Lindgren and Birdsall, 1970; Keen, 1970; Simonen, Chu, and Hendel, 1969; Keen and Aldridge, 1969; Parker and Thomassen, 1969). The feedback method commonly reported is shown in the upper portion of Fig. 1. We have used a somewhat more sophisticated method shown in the lower portion of Fig. 1 and have applied it to azimuthally propagating m = 1 mode "edge oscillations" (Müller, Corbin, and Palmer, 1970) in a single-ended cesium plasma Q-machine.

II. EXPERIMENTS

The sensor probe signal is fed into an oscilloscope which is triggered at the zero transition of each signal period. The oscilloscope gate signal is variably time delayed and triggers a single period, sine function generator tuned to a frequency slightly higher than the plasma oscillation frequency. The generator output, capacitively coupled to the suppressor probe, provides a constant-amplitude signal frequency- and phase-locked to the plasma oscillations. A slow square wave generator periodically switches

---

[*] On leave from Institut für Plasmaphysik, Garching, Germany.

on and off the trigger signal so that the feedback can be applied and
interrupted repetitively with any time sequence. The output from
this square wave generator is also the reference signal for a phase
lock amplifier which is used to measure the plasma signals with
high sensitivity.

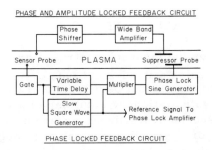

Fig. 1.   Feedback control methods

The Q-machine was operated at plasma densities $n_o \approx 10^{11} cm^{-3}$
and axial B-fields $\approx$ 2000 Gauss. The instability, with peak amplitudes near the periphery of the plasma column, could be controlled
by a d-c voltage, V_r, applied between the cesium injector ring
aperture (38 mm ID) and the tungsten ionizer plate. Fig. 2 shows
the variation in amplitude of the density oscillations, the oscillation
frequency, and the growth rates over a range V_r = -6 volts to +1 volt.
volt. The oscillations are noisiest and have largest amplitudes
when V_r is negative. With increasing V_r, the oscillations become
coherent. Simultaneously, their frequency begins to increase with
V_r and their amplitude to decrease. The growth rates (Müller,
Friz, and Palmer, 1970) increase with plasma density and with
decreasing V_r. The plasma is stable against this instability below
densities of about $1 \times 10^{10} cm^{-3}$ and can be stabilized at higher
densities for a sufficiently high V_r.

III. RESULTS

Figures 3 and 4 show the results of applying our feedback
control method. Figure 3 shows oscillation amplitudes, density
at the column axis, and frequency changes for various signal levels
applied to the suppressor probe as a function of phase shift between
the sensor and suppressor probe signals. Figure 4 shows oscillation
amplitude and density changes at phase shifts for maximum amplification and suppression as a function of feedback signal amplitude.
Various oscillation frequencies were obtained by changing V_r which
in turn affects growth rate and oscillation amplitudes (see Fig. 2).

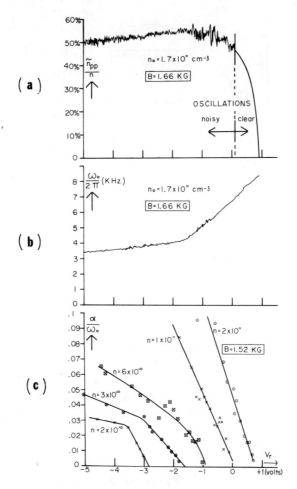

Fig. 2.   Edge oscillation amplitudes (top), frequencies (center), and growth rates (bottom) as a function of d-c voltage, V_r, applied between the cesium injector ring aperture and the tungsten ionizer plate.

Figure 5 shows normalized plasma density changes versus oscillation amplitudes varied by feedback amplification and suppression. The various curves differ in frequency, natural amplitude, and and growth rate.   The zero line corresponds to the d-c densities of the different oscillation-states with no feedback control applied. These preliminary measurements show that the axial plasma density n_o decreases with the squared oscillation amplitude $\tilde{n}^2$. This result suggests an $\tilde{E}_\theta \times B$ loss mechanism ($\tilde{E}_\theta$ = the azimuthal a-c electric field of the wave) according to a radial plasma flux equation

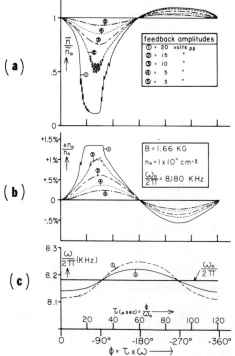

Fig. 3. Variation in oscillation amplitudes, (a) density at the column axis (b), and frequency (c) for different feedback signal amplitudes vs phase shift between sensor and suppressor signals.

$$j_r = -(c/2B) \ k_\perp \tilde{\phi} \tilde{n} \ \sin \beta \tag{1}$$

where it is assumed that the amplitude of the potential oscillation $\tilde{\phi}$ (which was not measured) varies linearly with the density oscillation $\tilde{n}$ and that the phase angle β between $\tilde{n}$ and $\tilde{\phi}$ does not vary with the instability amplitude.

Figure 6 shows typical radial profiles of the d-c plasma density, oscillation amplitude, and density disturbance caused by feedback at optimal amplification and suppression. Largest density deviations occur at radial positions where the oscillations have maximum amplitude. The radial density profile flattens with growing instability amplitude.

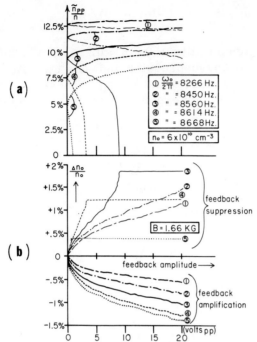

Fig. 4. Variation in oscillation amplitudes (a) and density at column axis (b) at phase shifts for maximum amplification/ suppression vs. feedback amplitude.

IV. INTERPRETATION

Our feedback method can be described by the stationary solutions of a second order linear differential equation

$$\ddot{n}_1 - (\alpha - \beta a^2/4)\dot{n}_1 + \omega_o^2 n_1 = \omega_o^2 n_2 \tag{2}$$

where $n_1 = a \sin \omega t$ is the plasma density oscillation, α is the linear growth rate and β the non-linear saturation coefficient, and where $n_2 = b \sin (\omega t + \phi)$ is a driving signal, with amplitude b and phase angle ϕ, which describes our feedback suppressor probe signal.

The stationary solutions of Eq. (2) are:

$$(a/a_o) (1 - a^2/a_o^2) = -\omega_o^2 b(\sin \phi)/\alpha \omega a_o \tag{3}$$

$$\omega^2/\omega_o^2 \;=\; 1 - b\,(\cos\phi)/a \qquad\qquad (4)$$

where a_o is the oscillation amplitude for $b = 0$. At $\phi = \pm 90°$, we have maximum amplitication/suppression and $\omega = \omega_0$. At $\phi = 0°$ and $-180°$, $\Delta\omega = \omega - \omega_o$ is largest and $a = a_o$. These results are in qualitative agreement with our experimental results shown in Figs. 3 and 4.

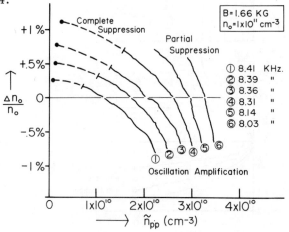

Fig. 5. Density changes on axis for various frequencies vs oscillation amplitude, which is varied by feedback amplification/ suppression. Zero corresponds to d-c densities without feedback.

In Fig. 7 (top) we have plotted the normalized oscillation amplitude for $\phi = \pm 90°$ according to Eq. (3). Figure 7 (center) shows experimental amplitude curves versus feedback amplitude for various oscillation states under condition of maximum oscillation suppression ($\phi = -90°$). In the states for which complete suppression can be achieved, oscillations suddenly collapse at a fixed critical value of $(n/n_o)_{crit.}$ which agrees well with $(a/a_o)_{crit.} = 1/\sqrt{3} = 0.577$, the point at which the two branches of the double valued amplitude curve converge (Fig. 7 - top). At this point $(\omega_o b/\alpha a_o)_{crit.} = 2\sqrt{3}/9 = 0.385$. Since we know the measured growth rates α/ω_o, the effective internal feedback signal amplitude b_{int} can be calculated and compared with the driving signal amplitude b_{ext}. From Fig. 7 (bottom), we see that the internal feedback signal b_{int}/a_o is not linear with b_{ext}, the suppressor probe signal. This can be understood from the non-linear characteristic of the suppressor probe.

Figure 8 shows measured changes in frequency versus suppressor probe signal for $\phi = 0°$. No change in oscillation amplitude is

measured, in agreement with Eq. (3).   The effective internal feed
back signal b_{int}/a_o derived from Eq. (4) is again non-linear with
respect to the applied signal.

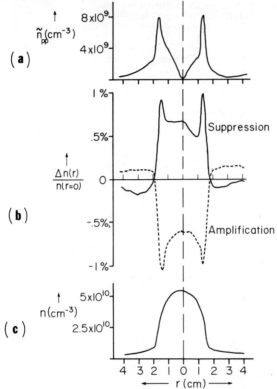

Fig. 6.   Typical radial profiles of amplitude (a) density change
due to optimal feedback (b) and density (c).

Growth rates can also be determined by our feedback method
according to

$$(\alpha/\omega_o)_F = \pm (1/2) \ (db/da) \big|_{b \to o} \tag{5}$$

derived from Eq. (3) for $\phi = +90°$ (+sign) and $\phi = -90°$ (-sign).
Using a phase sensitive method, $(\alpha/\omega_o)_F$ could be measured from
the amplitude changes at feedback on and off for very small
suppressor probe signals and compared with absolute values of
$(\alpha/\omega_o)_G$ obtained by exciting oscillations from the quiescent
plasma state (Müller, Friz and Palmer, 1970).   Over a wide range
of growth rates it was found that $(\alpha/\omega_o)_F$ is nearly proportional

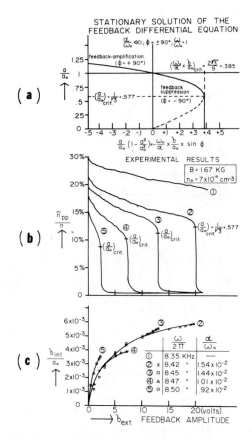

Fig. 7. Stationary solution of the feedback differential equation (a) . The portion to the right of the vertical zero line for optimal feedback suppression shows $(a/a_o)_{crit.} = 1/\sqrt{3}$. Experimental results (b) of oscillation amplitude versus feedback signal amplitude confirm this value. The effective internal feedback signal amplitude, b_{int}, is compared with the external feedback signal amplitude (c)

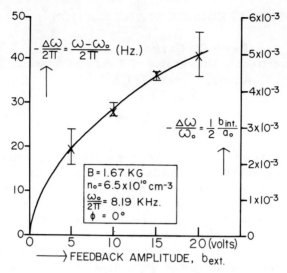

Fig. 8.   Measured changes in frequency (left vertical scale) and calculated values (right vertical scale) versus feedback signal amplitude.

to $(a/\omega_o)_G$.   From the slope of the $(a/\omega_o)_F$ versus $(a/\omega_o)_G$ curve. The optimal a-c coupling between suppressor probe and plasma can be calculated.

3.7 PASSIVE FEEDBACK STABILIZATION[*]

David C. Carlyle

Physics Department, University of Texas at Austin,
Austin, Texas 78712

ABSTRACT

A single-ended Q-machine has been modified to provide for the study of generalized boundary conditions. The sections of the segmented boundary can be interconnected through passive components to provide passive feedback. The ion density fluctuation amplitude of the Kelvin-Helmholtz instability , for densities of $\sim 10^8$ - 10^9 cm^{-3} and magnetic fields of $\sim$1-2 kG, decreased by 20-40% when feedback is applied. The DC peak density is increased by $\sim$10-30% upon stabilization. The effect shows a definite dependence on the capacitance and resistance in the passive circuitry.

I. INTRODUCTION

Plasma losses due to drift and flute modes have been successfully reduced by a number of researchers using feedback methods (Arsenin, Likhtenstein,and Chuyanov, 1968; Arsenin, Zhiltsov, and Chuyanov, 1969; Simonen, Chu, and Hendel, 1969; Hendel, Chu, Perkins, and Simonen, 1970; Keen and Aldridge, 1969; Keen, 1970; Parker and Thomassen, 1969). Lindman (1967,) has suggested that passive stabilization could be utilized for the same purpose. We have conducted an experiment to test that prediction.

II. EXPERIMENTS

Figure 1 shows the small single-ended cesium Q-machine (Rynn, 1964) used. Data were taken for densities n $\sim$10^8 - 10^9 cm^{-3} and for a uniform magnetic field $\underline{B}$$\sim$ 1 - 2 kG. The plasma is produced by contact ionization of a neutral cesium beam illuminating an electron-bombardment-heated 4.6 cm diameter tantalum plate. The plasma radius (r_p = 13.5 mm) is defined by a grounded aperature limiter. The plasma length is 56 cm. The conducting terminator is allowed to float, or is biased to reflect electrons. Results show no dependence on terminator bias. The plasma is collisionless.

[*]This work supported by Texas Atomic Energy Research Foundation and U.S. Army Research Office (Durham).

Two ion-biased spherical probes monitor the plasma.   The mid-column probe is moveable radially.   The end probe can be rotated about an offset axis.   Frequency and amplitude measurements are made with a Tektronix 1A7 preamplifier, a 1L5 spectrum analyzer plug-in, and with a Hewlett-Packard 302A wave analyzer.

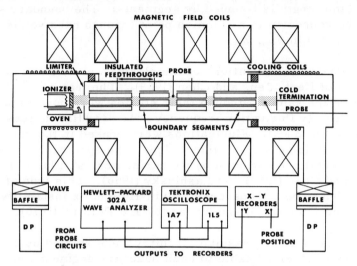

Fig. 1.   Schematic of the experiment showing axial arrangement of boundary segments.

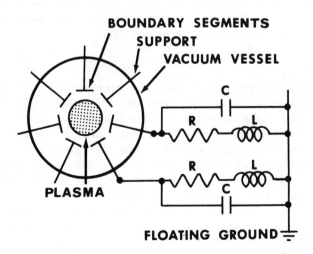

Fig. 2.   End view of the center section of the tube.   The boundary segments are shown.   Two typical circuits are shown, there are seven identical circuits around the circumference.

The center section of the tube has twenty-eight boundary segments: seven azimuthally (Fig. 2) with four axially at each azimuthal
position. The four axial segments are connected together, forming
seven segments parallel to $\underset{\sim}{B}$. The ratio of r_p to the radius of the
boundary segments is $(r_p / r_b) = 0.59$. Approximately 60% of the
plasma column length is bounded by segments. The boundary segments can be connected to a common point through passive element
circuits as shown in Fig. 2.

The experimental conditions of the plasma were chosen to match
closely those of Kent, Jen, and Chen (1969). The features of the instability were the same, although their studies were done in potassium. Ion-density fluctuation frequency (~12 - 23 kHz) increased with
decreasing $\underset{\sim}{B}$(~1 - 2 kG). Fluctuations were peaked in the region of
large electric field near the plasma edge. The spectrum shows a
large fundamental accompanied by its harmonics, similar to those in
Kent, Jen, and Chen (1969) and Lashinsky (1966). This type of wave
has been shown to be a transverse Kelvin-Helmholtz instability by
Kent, Jen, and Chen (1969) and by Jassby and Perkins (1970).

The effects of adding passive circuitry were studied by connecting 10 meters of shielded cable to each boundary segment and connecting them to a common ground point through circuits containing
1.5 MΩ variable resistors in parallel with inductors L variable from
30 mH to 105 mH. The length of the cables removes the inductors
from the effects of $\underset{\sim}{B}$. The cables also provide ~3nF parallel capacitance C to ground in each of the seven circuits.

III. RESULTS

When the floating ground is made the machine ground and all
seven resistors are at 1.5 MΩ no effect on the wave is observed.
As the resistances are changed to their minimum values of ~200 Ω,
the amplitude of the ion fluctuations is reduced in a continuous
manner. If the resistors are removed and the cables shorted to
ground (capacitance only) stabilization is still observed, although not
to the same degree as when a small resistance is in the circuit.
Apparently a small R is necessary for optimum stabilization. The
necessity of C is shown by removing the cables at the machine and
shorting the boundary segments directly to the vacuum vessel. In
this case no effect on the wave is observed. Removal or variation of
L over the range available produces no change in the fluctuations.
Apparently a CR network is necessary for stabilization of the Kelvin-
Helmholtz modes.

Typical results are, for $B \sim 1.5$ kG and $T \sim 2300°$ K:

$n(cm^{-3})$	Ion Density Increase (%)	Ion Fluctuation Decrease (%)
$\sim 5 \times 10^8$ (end)	20 (end)	35 (center)
$\sim 8 \times 10^9$ (end)	28 (end)	25 (end)

The measurements with the end probe are taken about 2 cm from the cold end of the boundary segments.

The boundary must be in electrical contact with the plasma through the ions. Biasing the plates to reflect ions turns off stabilization and restores the wave to its natural amplitude. Biasing to collect ions produces stabilization, but to no greater degree than connecting to ground without bias. No stabilization has been observed for a completely floating ground.

The density is increased across the entire radial profile. Radial floating probe potential curves show little change, indicating that we have not drastically perturbed the plasma.

Stabilization is observed throughout the range of B investigated. The effect is reduced at the extremes of B (~ 1 kG, 2 kG). This is apparently due to the parameters of the feedback network, and for these regions better results might be obtained by a different choice of circuit parameters. The frequency of the ion fluctuations at the extremes of B are just outside these frequency limits. Other factors may also be involved in this reduced effect, including changing sheath conditions and changing radial confinement of the plasma with B.

Other indications that a feedback mechanism is operating are provided by the observation that ion fluctuations can be made to grow at low magnetic fields. This effect is expected and is easily demonstrated in dynamic systems when positive feedback is applied. The major reduction in amplitude of ion fluctuations is produced by changing the resistance on sets of electrodes having a fixed separation. Selective mode damping has also been observed when two modes are present during changes in B. We have made no attempt to measure mode numbers.

IV. CONCLUSIONS

We have demonstrated that the amplitude of Kelvin-Helmholtz instability was reduced and plasma losses decreased by the application of passive feedback. It is to be expected that a system offering greater control over the circuit parameters, especially C, would result in optimum stabilization.

3.8 FEEDBACK STABILIZATION OF THE TRANSVERSE KELVIN-HELMHOLTZ INSTABILITY —— EXPERIMENT AND THEORY

T. K. Chu, H. W. Hendel, D. L. Jassby, and T. C. Simonen
Princeton University, Princeton, New Jersey 08540

ABSTRACT

By modulating the electron current flow to an electrostatic probe immersed in the plasma, the Kelvin-Helmholtz instability driven primarily by transverse velocity shear at the edge of a Q-machine plasma column is feedback stabilized. The measured feedback gain and phase delay and the oscillation frequency shift are in reasonable agreement with predictions from linear-theory quadratic-form representations constructed from the appropriate radial wave equation. Upon stabilization, the plasma density in the plasma interior increases.

I. INTRODUCTION

In the region of large electric field at the edge of the Q-machine plasma column there is a low frequency instability whose driving mechanisms have recently been identified as primarily $\underline{E} \times \underline{B}$ velocity shear and secondarily centrifugal force (Kent, Jen, and Chen, 1969; Jassby and Perkins, 1970). Hence it may be called a Kelvin-Helmholtz instability (Chandrasekhar, 1961). In this paper we describe an experiment in which this instability is stabilized by a feedback controlled electron sink, and compare the measured feedback parameters with values derived from linear fluid theory.

II. THEORY

The instability is detected by an electrostatic probe immersed in the plasma. This signal is amplified, phase-shifted, and capacitively coupled to a second electrostatic probe, the suppressor, so that the electron current flow to this probe is modulated (Simonen, Chu, and Hendel, 1969). The suppressor can be represented in the electron continuity equation by a uniformly distributed source term:

$$\frac{\partial \tilde{n}}{\partial t} + \nabla \cdot (n\underline{v})^{(1)} = -\sigma\tilde{n} = -\left|\sigma\right|e^{i\theta}\tilde{n} \tag{1}$$

By subtracting this perturbed electron continuity equation from the perturbed ion continuity equation, one derives, in the notation of

Perkins and Jassby (1970), the following radial wave equation which describes the Kelvin-Helmholtz instability, including FLR and collisional effects (Jassby and Perkins, 1970; Perkins and Jassby, 1970):

$$\frac{d}{dr} \, nr^3 w^2 \frac{d\psi}{dr} - \frac{m^2-1}{r^2} \, nr^3 w^2 \psi + \omega^2 r^2 \frac{dn}{dr} \psi$$

$$+ \, 2i \sum \frac{nr^3}{a^2} \frac{(\omega_D + \omega_E - \omega + i\sigma)(\omega - \omega_E)}{\omega - \omega_E + i\Sigma + i\sigma}$$

$$- \, \frac{1}{4} \, i n a^2 \nu_i r \, (\omega_D + \omega - \omega_E) \nabla_\perp^4 (r\psi)$$

$$+ \, \sigma m \omega_{ci} r^2 \frac{dn}{dr} \frac{(\omega_D + i\Sigma)(\omega - \omega_E)}{\omega_D(\omega - \omega_E + i\sigma)} \psi = 0 \tag{2}$$

where

$$\psi(r) = \frac{mc}{rB} \frac{\tilde{\phi}(r)}{\omega - \omega_E} \, \exp \, (im\theta + ik_\parallel z - i\omega t) \tag{3}$$

$$w^2 = (\omega - \omega_E)(\omega - \omega_E + \omega_D) \tag{4}$$

$$\omega_E = -\frac{mcE}{rB} \, , \qquad \omega_D = -\frac{mcKT}{rBe} \frac{1}{n} \frac{dn}{dr} \tag{5}$$

$$a^2 = \frac{2KT}{m\omega_{ci}^2} \, , \qquad \Sigma = \frac{k_\parallel^2 kT}{m_e \nu_{ei}} \, , \tag{6}$$

and $\tilde{\phi}$ is the perturbed plasma potential. Collisional viscosity can be neglected for the present experimental conditions.

The effect of the electron source on plasma stability can be found by constructing a quadratic form (Perkins and Jassby, 1970). Multiplying Eq. (2) by $\psi*$, integrating over radial space, taking the large electric field limit $\Sigma \ll |\omega - \omega_E|$, and using a value of Σ averaged over the radial wave domain, we obtain a quadratic equation in ω, from which the following solution is obtained:

$$\omega = \frac{1}{2}\left(2\bar{\omega}_E - \bar{\omega}_D - i\,\frac{2\Sigma}{k_\perp^2 a^2} \pm R^{\frac{1}{2}}\right) \tag{7}$$

where the most important terms of R are

$$R = \bar{\omega}_D^2 - 4\,\overline{(\omega_E - \bar{\omega}_E)^2} - \frac{8}{k_\perp^2}\left\langle\frac{1}{r_o^2}\right\rangle\bar{\omega}_E^2 + R_1 \tag{8}$$

where the first term represents FLR stabilization, the second velocity shear, the third centrifugal force, and

$$R_1 = -\sigma\,\frac{8A}{k_\perp^2 a^2}\left\langle\frac{(\omega_D + i\Sigma)(\omega - \omega_E)}{\omega - \omega_E + i\sigma}\right\rangle \tag{9}$$

The brackets and bars denote averages over $|\Psi|^2$ and $|\nabla\Psi|^2$, respectively. Without feedback, $R_1 = 0$; we then recover the results of Perkins and Jassby (1970).

A. Phase Delay. For instability, R must be negative. For optimum stabilization, the term representing the source R_1, must be real and positive. Thus we obtain the desired phase delay in different density regimes:

a) Low density: $\Sigma \gg \omega_D$, $\theta = \frac{\pi}{2} + \tan^{-1}\frac{\sigma}{\langle\omega - \omega_E\rangle}$ (10a)

b) Intermediate density: $\Sigma \approx \omega_D$, $\theta = \frac{3}{4}\pi + \tan^{1}\frac{\sigma}{\langle\omega - \omega_E\rangle}$ (10b)

c) High density: $\Sigma \ll \omega_D$, $\theta = \pi + \tan^{-1}\frac{\sigma}{\langle\omega - \omega_E\rangle}$ (10c)

B. Gain. At sufficiently large B with $R_1 = 0$, the velocity shear and centrifugal force are balanced by the stabilizing FLR and finite $k_\parallel$ effects. As B is lowered, the dominant change is in the velocity shear term. Using the empirical relation $E_{max} \propto B^{1/2}$, we obtain

$$\overline{\delta\, \omega_E^2} = \overline{(\omega_E - \bar{\omega}_E)^2} \approx \left(\frac{1}{4}\, \omega_{EM}\right)^2$$

$$\text{and} \quad \Delta\, \overline{\delta\omega_E^2} = \frac{1}{16}\, \omega_{EM}^2\, \frac{|\Delta B|}{B} \quad \text{if} \quad \frac{\Delta B}{B} \ll 1 \tag{11}$$

where ω_{EM} is the maximum value of ω_E.

For feedback stabilization, we require, from Eqs. (8) and (11),

$$|R_1| = \frac{1}{4}\, \omega_{EM}^2\, \frac{|\Delta B|}{B} \tag{12}$$

which reduces to

$$|\sigma| = \frac{1}{32}\, \omega_{EM}^2 \left[1 + (m^2 - 1)\left(\frac{a}{r}\right)^2\right] \frac{\left\langle \sqrt{1 + \left(\frac{}{\omega - \omega_E}\right)^2} \right\rangle}{\left\langle \sqrt{\omega_D^2 + \Sigma^2} \right\rangle} \cdot \frac{|\Delta B|}{B} \tag{13}$$

Generally, $\qquad \left\langle \left(\frac{\sigma}{\omega - \omega_E}\right)^2 \right\rangle \ll 1$.

C. <u>Frequency Shift.</u>   At optimum gain, as the phase delay, θ, is varied, there is a frequency shift. Expanding $R^{1/2}$, and noting that $-4\,(\overline{\delta\,\omega_E})^2$ is the dominant term in R, we obtain

$$\Delta\omega = \frac{-i\frac{1}{2}R_1}{\sqrt{4\delta\,\omega_E^2}}, \qquad \max \Delta\omega_r = \frac{1}{4}\, \omega_{EM}\, \frac{|\Delta B|}{B} \tag{14}$$

In the experiments, ω_{EM}/m is not negligible compared to ω_{ci}. Thus the ion zero-order polarization drift is important, and in Eqs. (13) and (14) ω_{EM} must be replaced by

$$\Omega_{EM} = \frac{1}{2}\, m\, \omega_{ci} \left\{-1 + \left[1 + \frac{4}{m}\, \frac{(\omega_{EM} - \omega_{DM})}{\omega_{ci}}\right]^{\frac{1}{2}}\right\} \tag{15}$$

III. EXPERIMENT

The experimental arrangement is that described in Simonen, Chu, and Hendel (1969), except that the detector and suppressor probes are located in the region of maximum electric field (and hence maximum wave amplitude) at the edge of the plasma column, as shown in Fig. 1.

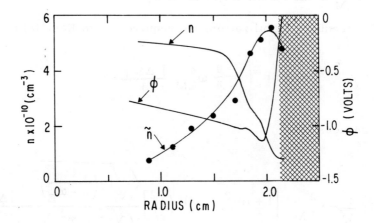

Fig. 1. Radial profiles of density fluctuation amplitude and equilibrium density and potential. The maximum value of $\tilde{n}/n$ was typically 5%. The suppressor probe is not effective if it is placed in the cross-hatched region.

Figure 2 shows the instability spectrum when an m = 2 mode is dominant, with and without feedback stabilization.

The input current to the suppressor probe was measured directly and displayed on a double-beam oscilloscope together with the detector ion saturation fluctuation signal, so that the phase delay, θ, could be directly measured. Figure 3 shows the m = 3 instability amplitude as a function of feedback phase delay, at constant feedback gain. Stabilization occurs at $\theta \approx 130°$, in reasonable agreement with the theoretical value of 160°. Upon stabilization the density inside the plasma column increases by a few percent. Maximum frequency shift was generally obtained when the phase delay was about 90° from the optimum stabilization value; however, the frequency shift was about 40 to 50% of the theoretical value calculated from Eqs. (11) and (15).

Figure 4 shows the experimental and theoretical feedback gain as a function of $\Delta B/B$ for an m = 2 mode. The instability amplitude for $\Delta B/B < 0$ appears to be due to enhanced thermal fluctuations (Kadomtsev, 1965; Chu, Hendel, Motley, Perkins, Politzer, Stix, and von Goeler, 1969) so we regard the actual onset as occurring at $\Delta B = 0$. The theoretical gain calculated from Eq. (13) is in good agreement with the measured values when the polarization correction of Eq. (15) is included.

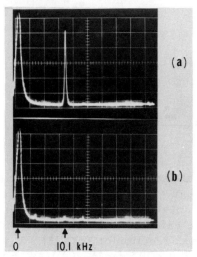

Fig. 2. (a) Frequency spectrum showing an m = 2 mode at 10.1 kHz and $|\Delta B|/B = 0.05$. The zero-kHz marker is instrumental. The range of the logarithmic scale is 40 db. (b) With feedback stabilization, the same mode is suppressed by 35 db or 98% in amplitude.

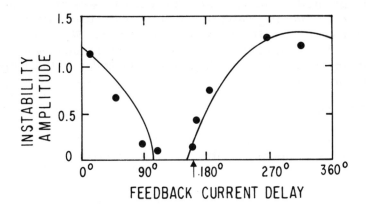

Fig. 3. Density fluctuation amplitude versus feedback phase delay for an m = 3 mode in a potassium plasma. B = 1.38 kG, $\langle \omega_D \rangle = 1.3 \times 10^5$ sec^{-1}, $\langle \Sigma \rangle = 4.2 \times 10^4$ sec^{-1}.

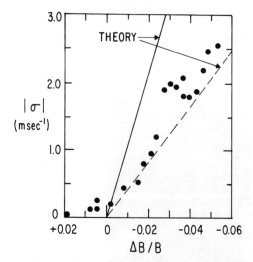

Fig. 4. Experimental and theoretical feedback gain, $|\sigma|$, versus $\Delta B/B$ for an $m = 2$ mode in a cesium plasma. The experimental points are determined from the measured suppressor probe currents. The theoretical gains are calculated from Eq. (13); the dashed line includes the polarization correction of Eq. (15). $B = 2.4$ kG, $\omega_{EM} = 3.6 \times 10^5$ sec^{-1}, $\langle \omega_D \rangle = 3.0 \times 10^4$ sec^{-1}, $\langle \Sigma \rangle = 2.5 \times 10^4$ sec^{-1}.

IV. CONCLUSIONS

The Kelvin-Helmholtz instability at the edge of the Q-machine plasma column has been stabilized by a feedback-controlled electron source. The feedback parameters are in reasonable agreement with the predictions of quadratic-form analysis using the linearized fluid theory for the Kelvin-Helmholtz instability, although the maximum frequency shift is less than predicted. We note that the theory should be applicable to Kelvin-Helmholtz modes which may occur in any plasma device which has large shear in the $\underline{E} \times \underline{B}$ rotation at the plasma edge.

ACKNOWLEDGMENTS

"This work was performed under the auspices of the U. S. Atomic Energy Commission, Contract No. AT(30-1)-1238."

3.9 FEEDBACK EXPERIMENTS ON MULTIMODE IONIZATION WAVES

A. Garscadden and P. Bletzinger
Aerospace Research Laboratories
Wright Patterson AFB, Ohio 45433

ABSTRACT

The influence of feedback and external driving signals on growth and saturation of ionization waves in a finite length positive column discharge has been investigated. Inserting a frequency-swept driving signal the ionization wave response exhibits eigenmodes which are equally spaced on the wavenumber axis. In the absence of any external perturbation only one mode is spontaneously excited. By external feedback the dominant ionization wave can be caused to mode jump. It is observed that with external electrodes, the mode changes occurred from the cathode, and propagated with a finite velocity (approximately that of the group velocity of the waves). Measurements of the spatial gain and its saturation for different discharge currents (or gain) are compared with the Landau saturation formula to give gain and saturation coefficients.

I. INTRODUCTION

The striating positive column of a low-pressure gas discharge is an example of a bounded plasma with several possible modes of oscillations. These oscillations can be observed optically and can be influenced by external longitudinal electric fields, their growth rates can be conveniently varied with discharge current. Their linear-theory dispersion relation has been derived and checked experimentally for a number of atomic gases. Thus, they can be effectively used to study mode coupling, feedback control, and time and spatial dispersion of spectra produced by nonlinear multimode oscillations.

II. THEORY

The properties of ionization waves or moving striations near the onset of the instability in glow discharges have been described by the Pekarek equation (Pekarek, 1968).

$$\frac{\partial n}{\partial t} = D \frac{\partial^2 n}{\partial z^2} + c_1 n - c_2 \int_z^\infty [\exp-a(\phi-z)]n(\phi,t)d\phi. \qquad (1)$$

The model and the notation for Eq. (1) are given by Lee, Bletzinger, and Garscadden (1966). This equation may be solved by a Fourier expansion

$$n(z,t) = Re \sum_{m=1}^\infty C_m(t)e^{ik_m z}, \quad k_m = \frac{m\pi}{L} \qquad (2)$$

and leads to the dispersion relation shown in Fig. 1. It is necessary

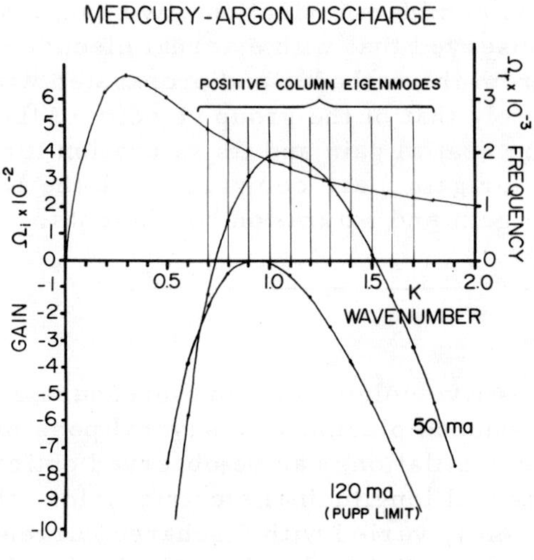

Fig. 1. Dispersion relation for ionization waves. 120 mA, $D = 663$, $c_1 = 1219$, $c_2 = 2053$, $a = 0.3$, 50 mA, $D = 714$, $c_1 = 2100$, $c_2 = 3630$, $a = 0.3$, Ω_r, shown for the Pupp limit, scales linearly by the factor c_2 for other currents.

to match the boundary conditions at both ends of the positive column. Apart from a possible phase factor, this leads to the condition that an integral number of striations is required, hence the positive column eigenmodes. The attenuation coefficient

$$-\Omega_i = \phi = Dk^2 - c_1 + \frac{c_2 a}{a^2+k^2} \qquad (3)$$

was calculated for a mercury-argon discharge for the Pupp limit, the onset of the striations (120 mA in our discharge). This linearized theory has also been used to calculate ϕ for 50 mA, where the observed striations have large amplitudes. Near onset, only the mode closest to the maximum of the gain curve is expected. However, under finite-gain conditions the theory predicts that the number of excited modes is given by

$$M \approx L((P+(P^2-4Q)^{\frac{1}{2}})^{\frac{1}{2}}-(P-(P^2-4Q)^{\frac{1}{2}})^{\frac{1}{2}})/\pi \qquad (4)$$

where

$$P \equiv (c_1-a^2D)/2D, \quad Q \equiv (c_2a-c_1a^2)/2D$$

so that a long column could support several modes. Experimentally, only one mode is found to be excited at one time; depending on how the discharge current is reached mode jumps and hysteresis occur. This indicates that there exists strong mode coupling. A short column length was found to suppress oscillations even when $L \sim \lambda$.

III.   EXPERIMENTAL RESULTS

The mode structure of the discharge was checked experimentally by applying an external AC signal of variable frequency close to the cathode end. At sufficiently high driving amplitudes the frequency of the ionization waves is locked in. The discharge response, recorded by a photomultiplier near the anode, is shown in Fig. 2. The results apparently follow the profile of the calculated curve except for the important differences that only one mode is excited at a time and even at discharge currents giving high gain conditions only 4 or 5 modes were recorded. The dominant mode moves to a higher frequency at higher currents, as expected. The length of the discharge column varies with current which affects the eigenmode frequencies slightly.

By applying amplitude and phase-locked feedback signals from external electrodes (Fig. 3), mode coupling and discharge response to feedback signals were studied. Figures 4 and 5 show the effects of a feedback signal, causing enhancement and suppression of the striation amplitude, respectively. A time delay between the applied signal and the discharge response can be noted. Further, under certain conditions (Fig. 6), after the suppression feedback was turned off, the discharge displayed ringing and several mode changes

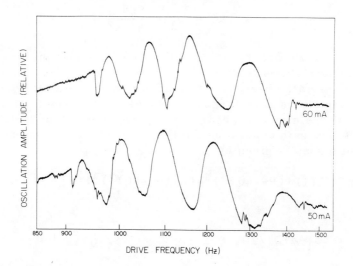

Fig. 2.　Photomultiplier record of discharge response to driving signal.

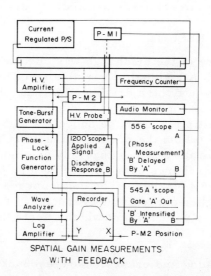

Fig. 3.　Circuit used to apply feedback and to measure discharge response.

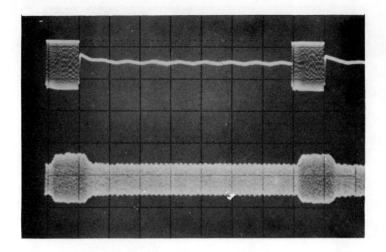

Fig. 4.   Enhancement of striation amplitude, upper trace
applied signal, direct feedback, lower trace, photomultiplier
record.   (20 msec/div).

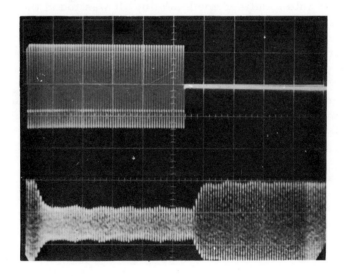

Fig. 5.   Suppression of striation amplitude, upper trace:
applied signal lower trace: photomultiplier record (10 msec/div),
89 mA phase-locked feedback.

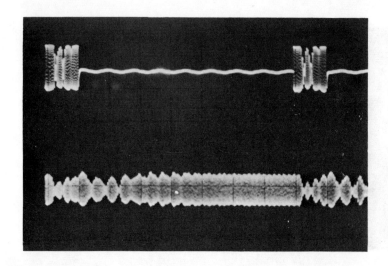

Fig. 6. Suppression of striation amplitude, upper trace: applied signal, direct feedback, lower trace: photomultiplier record (20 msec/div).

before it settled into a defined mode again. It was found that mode changes always propagated from the cathode and even with a balanced drive of the two external electrodes (no net AC field between electrodes and cathode) the time delay was independent of the electrode position from the cathode.

With these multiple mode conditions only a limited amount of suppression was possible. Since only single channel feedback was applied the discharge mode-switched with rapid growth rates on application of the feedback.

Growth rates of different modes can be estimated from these measurements (Fig. 5), but in the case of a steady-state, spatially growing and saturating mode, the gain and saturation coefficient can more easily be obtained from measurements of the oscillation amplitudes along the column (Sato 1970, Wong and Hai, 1969), Figs. 7 and 8. If the feedback signal was not exactly of the frequency of the particular mode, then for large feedback amplitudes a spatial amplitude beat occurred. At any given current, 4 or 5 amplitude maxima could be obtained with approximately the same gain coefficient, except for the highest-frequency mode which showed a lower coefficient. The dominant mode therefore

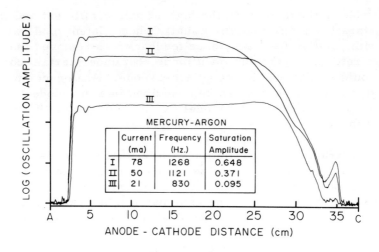

Fig. 7.   Spatial growth records showing exponential growth and saturation of ionization wave amplitude.

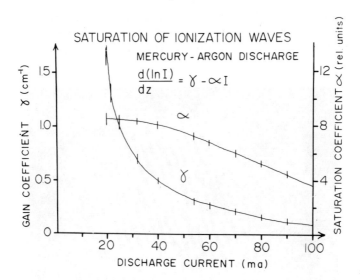

Fig. 8.   Gain and saturation coefficients vs discharge current.

corresponded to the mode with the highest gain coefficient and the lowest saturation coefficient α. Usually when a latent mode was excited with positive feedback, once feedback was removed the discharge returned to the dominant mode, but under certain conditions it would stay locked in the adjacent mode. Analog-computer results show these features can be described in a two mode phenomenological Van der Pol model (Keen, 1970) if amplitude-dependent mode-coupling terms are included.

The spectrum of the oscillation with feedback depended strongly on feedback amplitude and on the axial position, especially if the feedback frequency was not that of the dominant mode. The spectra observed were of the Bessel type (symmetric), indicating frequency modulation and also the single-side sideband spectrum (Abrams, Yadlowsky, and Lashinsky, 1969) indicating periodic frequency pulling. For instance, the single-side sideband spectrum can be obtained near the cathode but it will then disperse to the usual Bessel-function frequency-modulation spectrum towards the anode. At larger feedback amplitudes resulting in a correspondingly larger frequency modulation, the single-side sideband spectrum could be obtained throughout the discharge. A physical picture of the single-side sideband spectrum as produced by a travelling wave oscillation was obtained using a rotating mirror (space-time) display (Fig. 9).

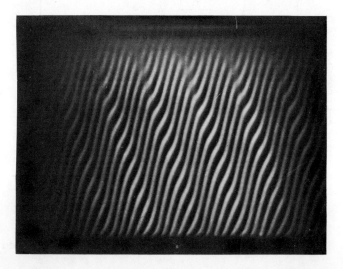

Fig. 9. Space-time display of sideband spectrum, this picture corresponds to lower sideband excitation. Discharge frequency 987 Hz. Applied signal 1356 Hz. Time increases downwards.

It can be noted that the generation of lower sidebands corresponds
to a periodic modulation (or rarefaction) of the ionization waves;
the upper sidebands were found to be generated by a compression
wave excited by an appropriate feedback signal of lower frequency
than that of the dominant mode. In addition to frequency dispersion
of the wave there also appeared to be a nonlinear amplitude
response of the discharge. With the method shown in Fig. 10, an

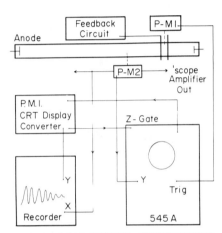

NON-LINEAR EFFECTS: INSTANTANEOUS
DISPLAY OF WAVELENGTH

Fig. 10. Circuit for instantaneous display of wavelength.

instantaneous display of the spatial wavelength could be obtained
(Fig. 11). It is noted that for constant frequency the wavelength is a
function of the oscillation amplitude. Expressed as an amplitude
dispersion this is shown in Fig. 12. When stabilized with positive
feedback, other modes show a similar behavior.

IV. SUMMARY AND CONCLUSIONS

Feedback has been used to investigate the properties of a
multimode travelling wave oscillation. The mode structure has
been found to correspond to the theoretical model, if boundary
conditions are applied. The dominance of a single mode can be
simulated in the Van der Pol model by the inclusion of a nonlinear
coupling term. By positive feedback, growth rates and saturation
parameters of latent modes could be determined experimentally.

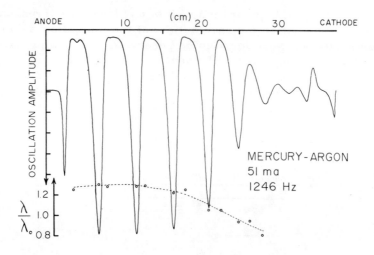

Fig. 11.   Oscillation amplitude and wavelength along discharge column.

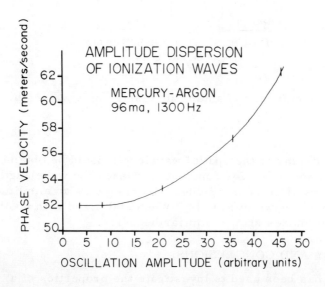

Fig. 12.   Amplitude dispersion of ionization waves.

Depending on feedback conditions, the oscillations show a frequency modulated or a combination of frequency and amplitude modulated spectrum which can disperse along the column. The specific wave forms in the plasma which produce these spectra have been identified. The oscillations have been shown to also have an amplitude dispersion.

The characteristic of strong amplitude and frequency dispersion mean that in an unregulated discharge, depending on the form of the initial perturbations, both pulse compression (chirp) and pulse spreading can occur even within the same wave packet causing splitting of the disturbance. This causes the discharge to appear noisy and turbulent, and obscures the eigenmode features. It is of interest to note that disturbance pulses whose phase velocity satisfies $\partial V_p / \partial f = - \partial V_p / \partial a$ where f frequency and a amplitude of the oscillation, should propagate with the minimum distortion. (Lighthill, 1965).

The use of negative feedback to quiet a discharge, or positive feedback to define a specified mode has application in the provision of quiet discharges for laser generation.

3.10 FEEDBACK STABILIZATION AND MEASUREMENT OF INSTABILITY COEFFICIENTS IN ELECTRON-HOLE PLASMAS

Betsy Ancker-Johnson and Hans Jacob Fossum
Boeing Scientific Research Laboratories and University
of Washington,* Seattle, Washington 98124

and

Alfred Y. Wong
University of California, Los Angeles, California 90024

ABSTRACT

More than 25db suppression of an m=0 (sausage) standing wave is obtained by feeding back only 0.5% of the instability oscillation output power. This mode is spontaneously generated at injected plasma currents of as little as~1A (corresponding to a plasma density of ~10^{14} cm^{-3}) in p-InSb with an applied axial magnetic field of ~400G. The linear growth rate is measured to be $\gamma = 8 \times 10^{6}$ sec^{-1}. The first and second nonlinear saturation coefficients are determined to be $\alpha = 3 \times 10^{10}$ sec^{-1} and $\beta = 0$. As either the current or magnetic field strength is increased, γ and α increase.

I. INTRODUCTION

The method for measuring linear growth rates and nonlinear saturation coefficients introduced by Wong and Hai (1969) is applied in this work to electron-hole plasmas. Such a plasma, produced by electrical injection into p-InSb as the plasma host, lends itself very well to these measurements since standing waves of single mode and frequency are spontaneously excited under steady-state conditions. The method is based on an experimental fit to the Landau amplitude equation, (Landau and Lifshitz, 1959), where $|A|$

$$\frac{1}{|A|^2} \frac{d|A|^2}{dt} = 2\gamma - \alpha|A|^2 - \beta|A|^4 + \sum_{n=3} C_n|A|^{2n} \quad , \quad (1)$$

is the wave amplitude, γ the linear growth rate, α and β the first

---

*Research in the Department of Electrical Engineering is partially supported by the National Science Foundation.

and second order nonlinear saturation coefficients, and C_n the higher order coefficients.

The m=0 (sausage) mode is now well understood in injected plasmas in p-InSb during moderately strong z-pinching (Chen and Ancker-Johnson, 1970). The spontaneous pinch oscillations observed in the absence of an applied magnetic field are caused by a standing sausage wave whose wavelength is twice the anode-to-cathode spacing. In the presence of magnetic fields applied within $\sim 25^{\circ}$ of coaxially with the plasma current, generally mixed m=0 and m$\geq$1 modes are observed with comparable amplitudes (Ancker-Johnson, 1964; Ancker-Johnson, 1970). Under coaxial alignment conditions an m=0 mode only is spontaneously excited at currents less than the critical current for pinching. The stabilization and, alternatively, enhancement of this mode is described, along with the first to our knowledge, measurements of growth rates and non-linear saturation coefficients of instabilities in solids. The m=0 mode in a moderately strong Z-pinch has also been stabilized by feedback and a subsequent paper will describe these results.

II. EXPERIMENTAL

A block diagram of the experimental arrangement is shown in Fig. 1. The single crystal of p-InSb is immersed in liquid nitrogen. The plasma is injected at the cathode contact and fills the sample long before the feedback measurements begin (Ancker-Johnson, 1968). When the direction of current flow is carefully aligned with the magnetic field direction, an m=0 mode is spontaneously excited whose oscillation amplitude is 2 orders of magnitude larger than the sum of the amplitudes of simultaneously occurring helical (m>0) modes. The operating range for this single mode is $\sim 0.8A \leq I_{plasma} \leq 1.3A$ and $\sim 370G \leq B_{\|} \leq 950G$. The saturated amplitude of the m=0 oscillations is conveniently controlled by adjusting either of these parameters. This current range lies just below the pinch threshold in the plasma under study.

The oscillations in voltage under essentially constant-current conditions are monitored by the small probes soldered on the sides of the sample. The center probe in the arrangement of Fig. 1 is used to sample the instability oscillations. This signal is fed through a 50db isolation circuit that includes a transformer. It is next amplified to minimize the noise introduced by the gate. Then the feedback signal is phase-shifted, gated, attenuated, and applied to the probe located nearer the cathode. A third probe located farthest downstream is used to observe the effects of feedback.

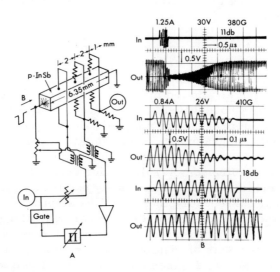

Fig. 1 - (A) Schematic diagram of the experimental apparatus. (B) Oscillogram showing suppression (top four) and enhancement (bottom two) of spontaneous m=0 oscillations. Operating conditions and scales are noted on the pictures. The db magnitudes refer to the attenuation in power between the "in" port and the surface of the sample.

The measured delay, caused by the circuitry, between the ports and oscilloscope traces marked "in" and "out" in Fig. 1 is $0.05\mu s$.

III. RESULTS

The first two oscillogram traces in Fig. 1 show the suppression (lower trace) obtained by feeding back five cycles of the instability oscillations (upper trace). Feeding back 2.5% of the instability power (observed at the "out" port), exactly out of phase, produces > 20db suppression. The next two oscillogram traces in this figure show, for a different set of operating conditions and expanded time scale, the essentially total suppression obtained with a feedback signal ~20db below the instability oscillation amplitude (2db comes from adjusting the amplitude of the feedback trace to be

the same as the saturated oscillation amplitude). The last two oscillogram traces show, with exactly the same operating conditions
as used for the two middle oscillogram traces except that the feedback signals are 180° out of phase, that the original oscillations
can also be enhanced as well as suppressed. The enhancement,
however, is only ~2.5db, whereas the suppression (middle oscillogram) is $\gtrsim$ 30db.

The oscillograms of Fig. 2 illustrate the ease with which feedback suppresses the instability oscillation when the operating conditions are relatively near threshold. For such operating conditions, (a) shows the instability amplitude (lower trace) with virtu-
ally no feedback.

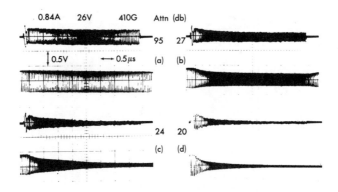

Fig. 2 - Four oscillograms illustrating the effectiveness of
feedback suppression. As in Fig. 1, the upper traces correspond
to the "in" signals and the lower traces to the "out" signals.

When a signal is fed back 30db below the original oscillation output
power, 5db suppression results as Fig. 2b shows. Suppression of
~20db is obtained when the feedback signal is 27db down, Figure
2c. In Fig. 2d no evidence of the instability oscillation is seen, and
this when the feedback signal is 23db down.

In these oscillograms as well as in the lower four traces in
Fig. 1, the effects of feedback are seen in the feedback signal itself. This is always the case if the feedback has a duration exceed-
ing the round trip delay, which is equivalent to five oscillation cycles in this experiment.

The data from such oscillograms, when reduced[*] to the form
of the left-hand side of Eq. (1), produce straight lines if the operating conditions are near threshold. Thus, $\beta=0$. As the operating
conditions penetrate the nonlinear region, the $|A|^{-2}(d|A|^2/dt)$
curves become concave. This suggests that at least one more coefficient beyond β is required to describe the nonlinear properties
of these plasmas.

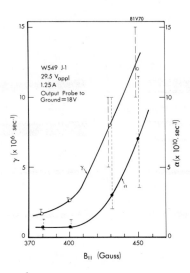

Fig. 3 - The linear growth rate γ and first nonlinear saturation coefficient α as functions of the applied magnetic field for a
fixed voltage a few volts above threshold.

[*] The normalization of the coefficients is made by comparing
the amplitude of the voltage oscillations to the average voltage level
at the point in the sample where the temporal measurement is made.

Magnitudes for γ and α are plotted in Fig. 3 as function of applied magnetic field for given voltage applied to the plasma. The large error bars for higher $B_\parallel$ are because the $|A|^{-2}(d|A|^2/dt)$ curves are concave. For magnitudes of the current even closer to threshold, the resulting γ and α as a function of $B_\parallel$ lie below those in Fig. 3. For example, with a smaller current and $B_\parallel = 410$G, $\gamma = 1.4 \times 10^6 \sec^{-1}$ and $\alpha = 0.9 \times 10^{10} \sec^{-1}$. With the same current but $B_\parallel = 460$G, $\gamma = 6 \times 10^6 \sec^{-1}$ and $\alpha = 3 \times 10^{10} \sec^{-1}$.

IV. CONCLUSIONS

Suppression of instability oscillations, generated by a current-carrying plasma in a magnetic field, by means of feedback is very effective in electron-hole plasmas. Little enhancement is obtainable, however, because strong nonlinear effects cause amplitude saturation. The linear growth rate and first two nonlinear saturation coefficients of an instability near its threshold in electron-hole plasmas have been measured. More coefficients are required to describe this instability's behavior well into the nonlinear regime. It is expected that such measurements can be readily made. It is also expected that similar information about spatially unstable modes (Sato, 1970) in electron-hole plasmas can be obtained.

CHAPTER 4

FEEDBACK EXPERIMENTS IN PINCHES, STELLARATORS, AND MIRRORS

4.1 OBSERVATION OF PLASMA DISPLACEMENT DUE TO DIPOLE CURRENT NEAR A HIGH-BETA PLASMA COLUMN

A. A. Newton, J. Junker,[*] and H. A. B. Bodin
UKAEA, Culham Laboratory, Abingdon, Berks., England

ABSTRACT

Plasma displacement due to dipole current near a high-beta plasma column is observed and interpreted. Its application to high-beta plasma feedback experiments is discussed.

I. INTRODUCTION

Feedback control of MHD instabilities is being pursued in conjunction with high beta toroidal experiments (Bodin and McCarten, 1969). The aim is suppression of slowly growing modes in systems with distributed axial and azimuthal magnetic fields which although initially stable may become unstable due, for example, to dissipation of currents in the plasma or stabilizing conducting shell. We consider systems in which the plasma is pinned by feedback applied magnetically by perturbing flux surfaces at one or more points around the torus. The present work describes observations of the effect on a plasma column of currents in a nearby coil.

II. EXPERIMENT

To design the feedback experiment it is necessary to know the magnitude and relative importance of the linear and quadratic contributions to the force (F_1 and F_2) produced by a current I_1 in a single sector coil of a long straight theta-pinch. Calculations for a straight sharp-boundaried plasma of area A_p and a small dipole of area A_d show that the linear force is

$$F_1(z) = 3A_p A_d I_o I_1 b(1 - 5Z^2/(b^2 + Z^2)) (b^2 + Z^2)^{-5/2} \qquad (1)$$

---

[*] On attachment from Institute für Plasmaphysik, Garching. Germany.

(b is distance of dipole from plasma axis, z the coordinate along the axis and I_o the plasma diamagnetic current per unit length). F_1 integrates to zero over an infinitely long column but locally exceeds F_2 except in regions around $z = b/2$ where F_1 changes sign. In the presence of a bulge F_1 no longer integrates to zero and is enhanced locally. For the above force with superimposed time variation i.e., $F_1 \propto \sin \Omega t \exp(-t/\tau)$ the plasma displacement, ξ, has been found by solving the wave equation with a forcing term.

Experiments have been carried out with a 2-m long uniform theta pinch (peak field 16 kG, risetime 9 μsec and 20 m Torr D_2) with alternating current in one sector coil (30 kA turns, $\Omega = 2 \times 10^6$ r sec^{-1}). Streak photographs, Fig. 1, show oscillatory motion in the plane through sector coil and plasma axes. Its direction reverses, as expected, with a linear force and its displacement is consistent with the calculation allowing corrections for the non-ideal geometry. Quadratic effects are not observed and the plasma column remains intact. For comparison the upper photographs was taken with no sector coil current. More details of these measurements are given elsewhere.[†]

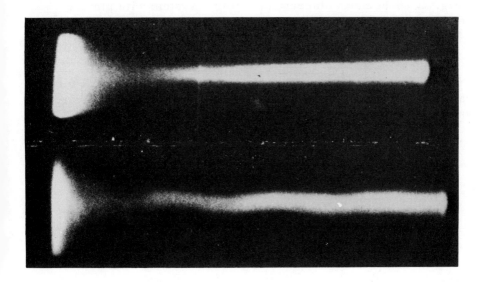

Fig. 1. Streak photographs of a theta pinch taken with (lower) and without (upper) one sector coil energized.

[†] J. Junker, A. A. Newton, and H. A. B. Bodin, submitted to Fourth European Conference on Controlled Fusion and Plasma Physics, Rome, 1970.

III. TOROIDAL GEOMETRY

It is possible to stabilize the m = 1 mode by restricting its amplitude at N stations around the torus (see Fig. 2).

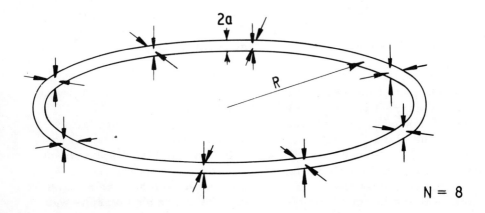

Fig. 2.   Schematic diagram of toroidal system with pinning forces.

The plasma is locally "pinned" by sets of orthogonal feedback forces and at intermediate positions the amplitude is restricted by hydro-magnetic coupling.

Using a simple model the essential features of systems of this type can be illustrated by an equation of the form

$$\frac{\partial^2 \xi}{\partial t^2} - V^2 \frac{\partial^2 \xi}{\partial z^2} - \gamma_0^2 \xi = \sum_{n=1}^{N} \alpha \, \delta(z - z_n)\xi \qquad (2)$$

in which the strength of the pinning forces is represented by the coefficients α and have a spatial form approximated by δ functions. Hydromagnetic connection is at the velocity V. The most dangerous mode (zero order axial mode) is suppressed if the number, N, of equally spaced stations exceeds $\pi \gamma_0 R / \{V \tan^{-1} (|\alpha| \, 2\gamma_0 V)\}$ in a torus of major radius R.

The feedback power, P_F, required is a strong function of growth rate. To keep the instability amplitude ξ less than a given value $\hat{\xi}$, an energy at least equal to the plasma kinetic energy must be provided at a rate faster than the growth rate. Thus for a gross mode $P_F \propto \gamma_o^3 \hat{\xi}^2 M$, where M is the plasma mass. Economic considerations based on this scaling restrict the fusion applications to slow growing modes.

IV. CONCLUSION

Measurements of the plasma displacement due to a dipole current near a theta pinch have yielded information on the linear force, to be used to test feedback with the m = 1 mode in a bulged theta and a screw pinch. Data obtained will enable feedback studies of gross modes in a toroidal device by pinning at one or more points.

Acknowledgments

It is a pleasure to acknowledge J.A. Wesson for useful discussions and C. A. Bunting and A. Wootton for experimental assistance. I.K. Pasco provided the theta pinch facility and helped with the plasma measurements.

4.2 FEEDBACK CONTROL OF COLLISIONAL DRIFT WAVES IN A TOROIDAL STELLARATOR*

C. W. Hartman, H. W. Hendel[†] R. H. Munger
Lawrence Radiation Laboratory, University of California
Livermore, California 94550

ABSTRACT

Identification of fluctuations and their suppression and enhancement by feedback are discussed for a plasma confined in a toroidal stellarator. Microwave produced Xe plasma having parameters $n = 10^{11} cm^{-3}$, $T_e \simeq 1 eV$ is studied in confining fields $B \simeq 1 kG$. A critical helical-winding current is observed above which the fluctuation level $\tilde{n}/n < 1\%$ and below which coherent fluctuations having $\tilde{n}/n \simeq 10\%$ are destabilized. The fluctuations are identified as the low azimuthal modes of ballooning-type drift waves. For currents smaller than threshold current, unfiltered feedback of appropriate phase suppresses the dominant lowest order mode while a higher order mode appears. For higher than threshold helical-winding currents excitation of oscillations by regenerative feedback to $\tilde{n}/n \simeq 10 - 20\%$ does not appreciably increase the rate of plasma loss under conditions $\tau_{confinement} \simeq 10 \ \tau_{Bohm}$.

I. INTRODUCTION

Plasma instabilities can appear as coherent modes particularly close to marginally unstable regimes (Hendel, Chu, and Politzer, (1968) where, because of their periodic nature and low growth rates, such modes lend themselves more easily to feedback control. This consideration led recently to a series of demonstrations of feedback stabilization (Arsenin, Zhil'tsov, and Chuyanov, 1969; Arsenin, Zhil'tsov Likhtenshtein, and Chuyanov, 1968; Parker and Thomassen, 1969; Keen and Aldridge, 1969; Simonen, Chu, and Hendel, 1969; Hendel, Chu, and Simonen, 1970) in linear devices. Moreover, improvement in plasma confinement with feedback stabilization of collisional drift waves has been observed, in agreement with that due to

---

*Work performed under the auspices of the U.S. Atomic Energy Commission.

[†] Permanent address: Plasma Physics Laboratory, Princeton, N.J. On leave from RCA Laboratories, Princeton, N.J.

transition from unstable to stable regimes by increase of ion gyroradius (Hendel, Chu, and Politzer, 1968).

We report experimental studies of feedback control of coherent drift waves in the hybritron device (Hartman, Munger, and Uman, 1969) operated as a toroidal stellarator. Unstable and stable regimes are found depending on the strength of the helical-winding current. In the unstable regime individual low-order azimuthal modes could be stabilized by feedback. In the stable regime excitation of large-amplitude modes by regenerative feedback was found to reduce the confinement time $\tau \simeq 10\ \tau_{Bohm}$ negligibly, confirming the existence of large nonfluctutional   losses.

II. APPARATUS

The experimental apparatus is shown in Fig. 1. The helical $\ell = 2$ and the toroidal-field windings were energized to provide

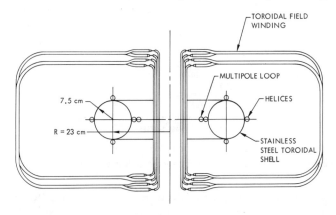

Fig. 1.   Schematic cross section of the hybritron.

confining magnetic fields $B \simeq 1\,kG$ and rotational transform $i \simeq \pi/4$ radians. No current was passed through the multipole loop. The Xe plasma was produced by applying for 0.1-1 msec 3.6 GHz microwave power up to 1.5 kW. Typical plasma parameters in the afterglow were $n = 10^{11}cm^{-3}$ and $T_e \simeq 1.0$ eV, and the confinement time was $\tau \simeq 2$-4 msec $\simeq 10$-20 τ_{Bohm}.

Instability identification and feedback studies were conducted with the array of single Langmuir probes and the feedback circuit

shown schematically in Fig. 2. Correlation of fluctuations around
the minor (poloidal) circumference of the plasma was established
with the seven-probe array spaced 1.5 cm, as indicated. Correlations in the toroidal direction were obtained by comparison of
signals from the seven-probe array, the tangential probe, and the
probes at $\phi = 0°$ and $90°$. The large-area suppressor probe acts
as an electron sink for stabilization of waves detected at other points
on the torus. The detected $\tilde{n}$-signals are phase shifted, amplified
by a broad-band amplifier, and fed back as indicated.

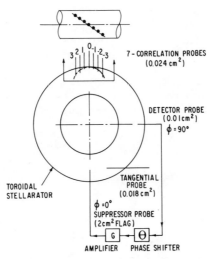

Fig. 2. Schematic illustration of toroidal stellarator, diagnostic
and feedback Langmuir probes, and feedback circuit.

Typical plasma decay and equilibrium distributions taken at
$t = t_0$ are shown in Fig. 3, for a Xe plasma. The relatively coherent, large-amplitude oscillation appears during the plasma decay
when the helical winding current $|I_H| \lesssim 1.5$ KA. For $|I_H| \gtrsim 1.5$ KA
fluctuations abruptly decrease to a level $\tilde{n}/n \lesssim 0.01$.

III. IDENTIFICATION OF FLUCTUATIONS

Identification of fluctuations as collisional drift waves is based
on correlation measurements establishing wavelengths $\lambda_\perp$, $\lambda_\parallel$,
phase velocity v_p, phase between $\tilde{n}$ and $\tilde{\phi}$, and wave localization.
Typical correlation measurements are shown in Fig. 4. For the
conditions shown the wave was found to propagate in the electron
diamagnetic-drift direction with $v_p \simeq 2.3 \times 10^4$ cm/sec. This
phase velocity is in good agreement with the electron diamagnetic
drift velocity, v_{de}, computed from measured T_e, $\nabla n/n$, and B,
where $v_{de} = ckT_e/eB (\nabla n/n) = 3.0 \times 10^4$ cm/sec, corrected

for a small electric field (plasma core positive) drift velocity $v_E \simeq -7.5 \times 10^3$ cm/sec. The phase difference between $\tilde{n}$ and $\tilde{\phi}$

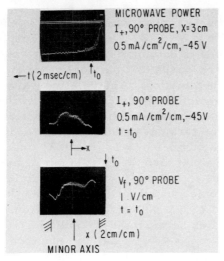

Fig. 3. Typical equilibrium properties for Xe plasma, I_T = 1.0 ka, I_H = 1.0 ka. At t = t_o, T_e = 0.9 eV; x is the radial coordinate measured from the minor axis of the vessel at the mid-plane.

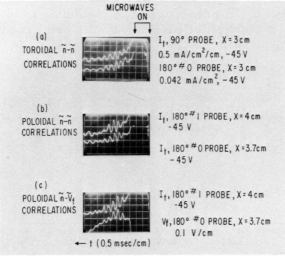

Fig. 4. Fluctuation correlations of $\tilde{n}$-$\tilde{n}$ in the toroidal dir-ection (a), $\tilde{n}$-$\tilde{n}$ in the poloidal direction (b), and $\tilde{n}$-$\tilde{\phi}$ in the poloidal direction, Xe plasma, I_T = 1.0 ka, I_H = 1.0 ka.

(the floating voltage $\tilde{V}_f$) obtained from Fig. 4 (b) and 4 (c) is 45° with $\tilde{n}$ leading $\tilde{V}_f$, consistent with wave growth.

In the regime of interest, near marginal I_H, low order poloidal modes are observed with f $\simeq$ 1, 2, 3 kHz. With conditions adjusted so that the f = 3 kHz mode is dominant the poloidal wavelength is $\lambda_p \simeq 7.8$ cm, which corresponds to a poloidal mode number $\ell = 3$, for the region of wave localization. The relative wave amplitude $\tilde{n}/n$ is nearly zero on the small major radius side of the plasma and close to 0.1 on the large major radius side, indicating a ballooning mode. The measured toroidal wavelength $\lambda_t \simeq 150$ cm corresponds to a toroidal mode number m = 1. From λ_p and λ_t we obtain wave numbers $k_p = 0.8$ cm^{-1} and $k_t = 0.015$ cm^{-1}, and from the flux surface localization, $k_\psi = 2$ cm^{-1}. When the field strength and rotational transform are accounted for, the wave has $k_\perp a_i \simeq 0.5$ (a_i evaluated at T_e) and $k_\| = 0.009$ cm^{-1} ($\lambda_\| = 117$ cm). The parallel phase velocity lies between the ion and electron thermal velocities i.e., $v_{ti} \ll v_{p\|} \ll v_{te}$, and the electron mean-free path is $\lambda_{mfp} \tilde{<} \lambda_\|$ indicating that the drift waves are collisional. Further identification of fluctuations with drift waves is based on comparable measured values of $\tilde{n}/n$ and $e\tilde{V}_f/kT_e$ and localization of the wave in the region of maximum density gradient. The source of ballooning has not been identified experimentally. It may be due to poloidal variation in the properties of the confining field ($\nabla \overline{B}$, shear, etc.) or to contact of the plasma with the inner conducting wall of the vacuum vessel.

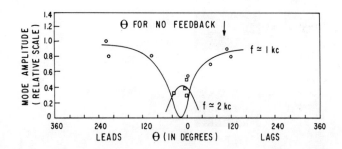

Fig. 5. Mode amplitude versus feedback phase Θ, Xe plasma, I_T = 0.98 ka, I_H = 1.24 ka.

IV. FEEDBACK CONTROL OF FLUCTUATIONS

In the unstable regime, specific modes could be feedback controlled. To avoid overloading the amplifier and disturbing the plasma during heating, the feedback circuit was gated on only during plasma decay and filtering was used to exclude the low-frequency oscillations associated with the decay. With I_H adjusted so that the $f \simeq 1$ kHz ($\ell=1$, m=1) mode was dominant without feedback, the mode amplitude shown in Fig. 5 was obtained as a function of feedback phase shift Θ for an amplifier gain which provided optimum stabilization. Maximum suppression of the $f = 1$ kHz mode occurs when $\Theta = 0°$, leading the no feedback value $\Theta = 120°$. This result is similar to that predicted by theory for linear geometry (Simonen, Chu, and Hendel, 1969). In the vicinity of $\Theta = 0°$, the next order mode at $f \simeq 2$ kHz ($\ell=2$) appears. Since broad-band feedback was used, it is not determined whether the $f \simeq 2$ kHz mode is excited by regenerative feedback or results from nonlinear wave interaction.

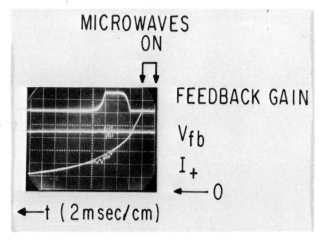

Fig. 6. Plasma decay with and without mode excitation by feedback, Xe plasma, $I_T = 0.98$ ka, $I_H = 1.71$ ka.

Measurements of plasma loss associated with the presence of drift waves were conducted in the stable regime $I_H \gtrsim 1.5$ ka. Drift waves (with correct $\tilde{n}$, $\tilde{\phi}$ phase) were excited by adjusting $\tilde{\phi}$ for regenerative feedback to produce a burst of oscillations with $\tilde{n}/n \simeq 0.1$ as shown in Fig. 6. The plasma density decay ($T_e \simeq$ constant) measured with a Langmuir probe is seen to be relatively unchanged by the excitation of coherent fluctuations, indicating a dominant non-fluctuating loss. This direct observation is in agreement with earlier indirect indications that fluctuation-induced losses can be small (relative to other losses) for certain conditions in toroidal systems (Hartman, Munger, and Uman, 1969; Anderson, Birdsall, Hartman, Lauer, and Furth, 1969).

V. CONCLUSIONS

Fluctuations in a Xe plasma confined in a toroidal stellarator have been identified as ballooning collisional drift waves. By feedback, selective stabilization of low-order modes has been achieved although complete stabilization was not observed. In regimes of quiescent plasma decay excitation of large amplitude drift waves by regenerative feedback was found to reduce plasma confinement negligibly, indicating that the dominant losses are not due to drift waves.

4.3 FEEDBACK EXPERIMENTS IN A HOT ELECTRON PLASMA[*]

Glenn Haste
Oak Ridge National Laboratory
Oak Ridge, Tennessee 37830

ABSTRACT

A plasma formed by electron cyclotron heating operates in one of two modes: A high density mode above a transition pressure ($\sim 5 \times 10^{-6}$) and a low density mode below. Below the transition pressure the plasma is unstable to flutes due to the electron drift. We have found that instability signals which are fed back to an electrode at the radial edge of the plasma alter the instability markedly. The frequency spectrum is reduced from a broad spectrum to a single frequency associated with self-induced oscillations. The fluctuations in electron density which are associated with the flute signals are eliminated. The mean density remains about at the same level when feedback is applied.

I. INTRODUCTION

A plasma formed by electron cyclotron resonance heating in the IMP Machine has been recently used for feedback experiments. In this device electrons are heated using 8 mm microwave power in a simple mirror configuration. The plasma characteristics depend on the neutral pressure as can be seen in Fig. 1. Above a transition pressure ($\sim 6 \times 10^{-6}$ Torr H_2) the hot electron density is 10^{11} and is independent of the pressure. The electron temperature, as determined from bremsstrahlung measurements is ~ 30 keV in this high density mode. Below the transition pressure the density is at least two orders of magnitude lower and varies exponentially with pressure. The electron temperature is not measured in this mode, but indirect evidence suggests that it is roughly the same. Flute instabilities are observed in this mode, and these flutes are subject, to a degree, to external control.

II. EXPERIMENT

The feedback experiments consisted of sensing the flute signals with one probe and applying the same signal, amplified without a

[*]Research sponsored by the U. S. Atomic Energy Commission under contract with the Union Carbide Corporation.

phase change, to an electrode on the opposite side of the plasma.

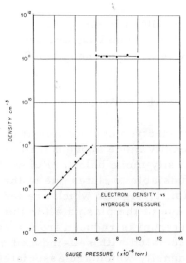

Fig. 1. Electron Density versus Hydrogen Pressure.

The electrode arrangement is shown in Fig. 2. The top electrodes
are used as sensors and serve to determine the flute mode number,
which is predominately m = 1.

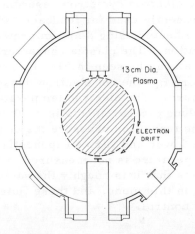

Fig. 2. IMP Electrode Arrangement.

The plasma density is monitored by measuring the ionization
current which flows along flux lines. This current fluctuates due to
the flute instability, with the flute activity preceding a reduction in
the ionization current.

III. RESULTS

The frequency spectrum for these flutes is shown in Fig. 3. Frequencies are seen which extend out to 300 kHz, with the largest

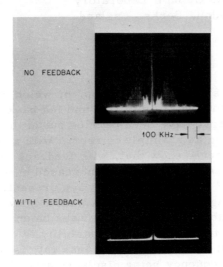

Fig. 3. Flute Frequency Spectrum.

signals occurring at 150-200 kHz. When using feedback with a voltage gain of 10^5 the spectrum is very different. There remains only a single frequency at 250 kHz. The average value of the ionization current remains essentially unchanged ($\pm 20\%$) but the fluctuations are very much smaller.

IV. DISCUSSION

It is apparent that the flutes limit the density in that they cause plasma losses. Thus, their elimination should result in an increased plasma density. Since the density does not increase, the remaining oscillation must lead to plasma losses as well. These oscillation-induced losses have yet to be measured.

4.4 FEEDBACK EXPERIMENTS ON A HIGH ENERGY PLASMA CONTAINED IN THE PHOENIX II SIMPLE MIRROR MACHINE

V. A. Chuyanov,[*] E. G. Murphy, D. R. Sweetman, and E. Thompson
UKAEA, Culham Laboratory
Abingdon, Berks, England

ABSTRACT

Experiments using a single feedback loop, by which electro-static signals from a plasma were amplified and fed back as potentials to the plasma boundary, have been carried out in the PHOENIX II mirror machine working as a simple mirror. With a flute unstable plasma the experiments show that, with feedback the density at which appreciable losses were observed increased by a factor of two. However, a residual instability was always present. Theoretical investigations, using a simple plasma model, show that the experimental plasma-feedback system will always be unstable and, in particular, the density threshold of the instability is lower with feedback than that without; the lower plasma loss rate is thought to be due to the instability frequency being displaced away from the magnetic drift frequency of the ions. Measurements of the effect of the feedback system on a normally stable plasma showed that the system was unstable at frequencies close to those predicted by theory. It is concluded that a simple electrostatic feedback system capacitatively coupled to the plasma would be insufficient to stabilize this reactive instability even if it had many feedback loops.

I. INTRODUCTION

The PHOENIX II machine, working as a simple mirror, has a monoenergetic plasma built up by Lorentz ionization of a 15-20 keV neutral hydrogen atom beam injected across the confining magnetic field which in the experiments described below was 11-12 kilogauss. The resulting plasma can be thought of as an 8 cm radius disc about 2 cm thick with a particle density determined by the equilibrium between the trapped beam current and the ion loss processes which, in the case of stable plasmas, is charge exchange with the background gas. However, at a density of $2\text{-}3 \times 10^8 cm^{-3}$, depending on the injection energy and the strength of the magnetic field, an

---

[*] On leave from the Institute of Atomic Energy (I. V. Kurchatov) Moscow, U.S.S.R.

m = 1 flute instability appears and further attempts to increase the density beyond this value are frustrated by the losses due to this instability.   The experiments described here were aimed at determining the effect of a feedback loop with a finite bandwidth on the flute-unstable plasma.

II.   EXPERIMENTS

The feedback system used is shown schematically in Fig. 1 and consisted of a sensing electrode capacitatively coupled to the plasma,

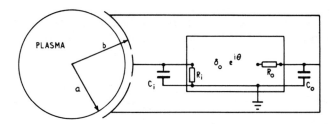

Fig. 1.   Schematic diagram of the feedback system.

amplifiers, and a means of shifting phase and control electrodes symmetrically placed around the sensing probe.   The whole electrode structure, covering a sector of about 80° of the circumference of the plasma, was placed 8.5 cm from the axis of the magnetic field and determined the plasma radius; this was taken to be one Larmor radius less than this distance.   The output stage of the amplifier system was capable of delivering 150 V peak-to-peak into the capacitative load used.   Phase shifts were obtained by using delay lines and also with a lock-in system incorporating a voltage-programmable filter.

The equations of Arsenin and Chuyanov (1968) indicate that there are two regions of stability, one when the potential fed back is such as to oppose the perturbed potential and the other when the applied potential is in the same sense as that created by the plasma.   The first region is a direct consequence of the sharp-boundary plasma model and it can be shown that this region of stability does not exist

for a plasma with diffuse boundary though the threshold density can
be increased to a new limit as the feedback loop gain is increased.
This regime was briefly investigated and an increase in density of
about a factor 2 was achieved due to, at least partial, suppression of
the flute; the plasma density was again limited by losses. The ;
second region of stability is, theoretically, attainable for plasmas
with diffuse boundaries and was studied extensively.

III. RESULTS

Earlier results had shown the effectiveness of such a feedback
system in decreasing plasma losses due to the flute instability.
Fig. 2 shows the effect of feedback on the plasma density for various
values of expected line density, i.e. density integrated along a flux
tube which would have resulted from the equilibrium between the
injected ions and charge exchange on the background gas. If no other
loss processes were present the plot would follow the dashed line.
It can be seen that without feedback the line density is limited by

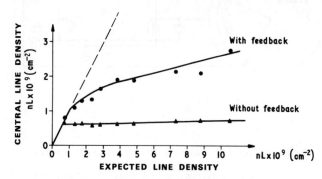

Fig. 2. Variation of central line density with that expected if
no losses are caused by instability showing the effect of feedback.
The dashed line indicates no losses other than by charge exchange.

flute losses to about $6 \times 10^8 cm^{-2}$. Feedback had a beneficial effect
in that an improvement of about a factor 4 in density was possible
albeit in the presence of large plasma losses.

Fig. 3 shows the effect of loop gain on the line density, frequency and amplitude of the low frequency oscillations for feedback
loops using wide and narrow band amplifiers interacting with a

highly unstable plasma. In the narrow band case a 1.46μs delay line in the circuit was found to give an optimum phase shift as far as density was concerned. The critical amplification δ^* above

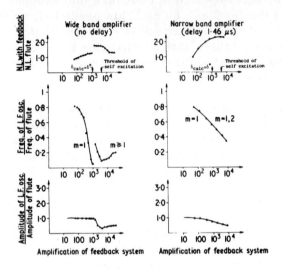

Fig. 3. Variation of line density, frequency and amplitude of the instability with amplifier gain for wide band and narrow band cases. δ^* is the critical amplification above which stability should have been achieved. m is the azimuthal mode number of the instability.

which the system should be stable according to the Arsenin and Chuyanov criterion is marked, as is the gain at which the feedback system oscillates in the absence of a plasma. For the case of the wide band amplifier it can be seen that for amplifications much less than the critical amplification the measured quantities vary slowly with amplification. As the amplification approaches the critical value there is a rapid fall in the observed oscillation frequency, though the azimuthal mode remained m = 1, and, as the amplification exceeds the critical one the plasma changes abruptly to a new state having a higher density and, simultaneously, a more complicated instability mode structure. The oscillations become very irregular and are a mixture of modes. Increasing the gain further resulted in decreasing the density and increasing the frequency and amplitude of the oscillations. In the case of the narrow band amplifier these features are in general also observed but the transitions are much smoothed out.

IV.   DISCUSSION

This behavior was quite at variance with what was expected on the basis of existing theory which did not take into account the frequency response of the feedback loop. This predicted, at least as far as the frequency of the oscillations is concerned, that the plasma should become more unstable, without changing the real part of the frequency, until the critical amplification is exceeded and then, discontinuously, should become stable.

The non-linear phenomena that determine the amplitude of the low frequency oscillations and govern the plasma losses are complicated and have yet to be fully investigated. However the increases in density are thought to have been achieved by changing the frequency of the unstable wave so as to decouple it from the plasma ions which are precessing at twice the frequency of the flute in the magnetic field gradient.

In order to study the behavior of the real part of the frequency, measurements of the frequency were made for various feedback loop gains and phase shifts and the results compared with a simple theoretical model consisting of a cylindrical plasma of radius a surrounded by a wall at radius b immersed in a magnetic field such that the magnetic precession frequency is independent of radius. The potential on the wall is assumed to be maintained at δ times the potential of the plasma surface at the same azimuth i.e. an infinite number of feedback loops. This model leads to the dispersion relation given by Arsenin and Chuyanov

$$\delta \left(\frac{a}{b}\right)^{|m|} = 1 + \frac{\omega_{pi}^2}{2\omega_{ci}^2} \left\{1 - \left(\frac{a}{b}\right)^{2|m|}\right\} \left\{1 + \frac{|m|\,\omega_{ci}\omega_D}{\omega(\omega+m\omega_D)}\right\}$$

where m is the azimuthal mode number, ω_{pi}, ω_{ci}, ω_D the plasma, ion cyclotron and magnetic drift frequencies respectively. With the feedback loop drawn in Fig. 1 it can be shown that

$$\delta = \delta_o\, e^{i\theta} \left\{\frac{i\,\omega\,\omega_H}{(\omega+i\omega_L)(\omega+i\omega_H)}\right\}$$

where $\omega_H = 1/C_o R_o$ and $\omega_L = 1/C_i R_i$.

Solutions of the equation with the proper frequency dependence showed that there was always at least one unstable root. Figure 4 shows a typical plot of the real part of the frequency against loop

gain for a stable plasma whose density is equal to the critical
density in the absence of feedback and for m = 1.   In these and subsequent plots the loop gain has been normalized to be independent

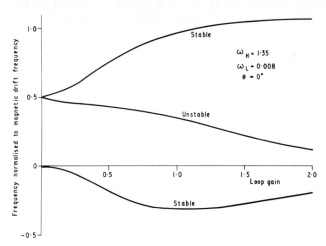

Fig. 4.   Typical solutions of the dispersion equation showing the
real frequency as a function of loop gain.   Loop gain has been normalized and is defined as

$$\text{Loop Gain} = \frac{\delta_o}{\left(\dfrac{b}{a}\right) + \left(\dfrac{b}{a} - \dfrac{a}{b}\right)\dfrac{\omega_{pi}^2}{2\omega_{ci}^2}}$$

The fourth root close to ω_L is not shown.

of geometrical factors and to eliminate a slight density dependence
and is re-defined as

$$\text{Loop Gain} = \frac{\delta_o}{\left(\dfrac{b}{a}\right) + \left(\dfrac{b}{a} - \dfrac{a}{b}\right)\dfrac{\omega_{pi}^2}{2\omega_{ci}^2}} \ .$$

It can be seen that there is one unstable root with a frequency
between 0 and 0.5.   The fourth root (not shown) seems always to
be stable and to be near ω_L.   At very high loop gains the two
stable roots can also become unstable but their growth rates are
always much lower than the main unstable root shown whose growth
rate is generally close to ω_H for significant loop gains.   It is
reasonable that there should be one unstable root because the flute

instability is a reactive instability involving coupling of a positive to a negative energy wave when the instability threshold is exceeded. Any phase shift in the feedback loop can cause the feedback loop to interact with either of these waves, depending on the direction of the phase shift and can cause an instability of the system.

Fig. 5 shows the experimentally obtained instability frequency plotted against loop gain normalized to the critical gain, also plotted is the theoretically calculated frequency taking into account the frequency response of the system used.   This has been done for a stable plasma with a density equal to the threshold for instability in the absence of feedback.   It can be seen that there is good agreement

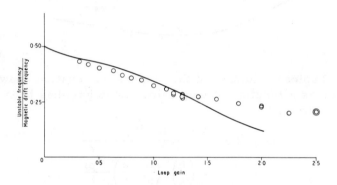

Fig. 5.   Comparison of experimental and theoretical values of the real part of the unstable frequency as a function of loop gain. Density = critical density, ω_H = 1.35, ω_L = 0.008, no phase shift.

especially at lower values of loop gain.   The values of ω_L and ω_H were taken to be those frequencies at which the phase shift was ± π/4 and the biggest uncertainty was in the estimate of δ_o which involved a measurement of the capacity between the plasma and sensing electrodes.

The frequency of the instability also varies with phase shift for a fixed loop gain and Fig. 6 shows this variation for a gain of 1.12. The theoretically calculated frequency is also plotted and it can be seen that the agreement is no more than reasonable.   At higher loop gains than this the discrepancy between experiment and theory becomes much greater and at a loop gain of 1.6 the frequency becomes

almost independent of phase shift, whereas theoretically near to
zero phase it should be very sensitive to phase shift.   The dis-
crepancies at high values of loop gain are thought to be due to stray
capacities not accounted for in the theoretical model; these are
such as to cause the feedback system to oscillate in the absence of
plasma at a loop gain of about 3.

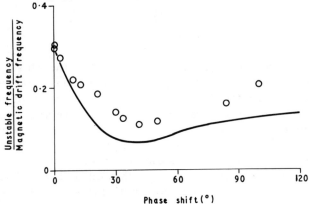

Fig. 6.   The effect of phase shift on the real part of the unstable
frequency.   The theoretical line is for density = critical density,
$\omega_H = 1.35$, $\omega_L = 0.008$, $\delta_0 = 1.12$.

Measurements of growth rates of the instability give values
which are an order of magnitude lower than those predicted by the
theory.   This may be a consequence of the fact that the precessional
drift frequency of the ions is not constant throughout the plasma thus
giving rise to some damping of the wave.

These experiments have shown that the frequency response of
the feedback loop is very important in determining the properties of
the plasma-feedback system and that the predictions of a very simplified theoretical model are reasonably well borne out by experiment.   If, as seems likely, the theoretical model is adequate then,
in contrast to the case of dissipative instabilities, there is no
region in which a capacitatively coupled system can stabilize this
reactive instability.

Future work will be aimed at determining whether it is possible
to devise a feedback system whose frequency response is such that
stability can be achieved.   Lashmore-Davies (1970) has shown that
not all feedback loops with frequency dependent phase shifts will be
unstable.

4.5 FEEDBACK STABILIZATION OF THE RAREFIED PLASMA CYCLOTRON INSTABILITY

V. A. Chuyanov, V. Kh. Likhtenshtein,
D. A. Panov, and V. A. Zhil'tsov
Atomic Energy I. V. Kurchatov Institute, Moscow, USSR

ABSTRACT

A surface of finite conductivity placed parallel to the plasma boundary can absorb enough energy to suppress the long-wavelength, unstable, low-increment oscillations. Feedback systems with regulated amplification and phase shift work like such a surface. Using such regulated feedback control as artificial damping, experiments were carried out in the OGRA-2 mirror machine. The results demonstrated the possibility of suppressing unstable cyclotron oscillations of different mode structure and showed that the broadening of the confined ion spectra is connected with the symmetrical mode of oscillation and the axial expansion of the plasma is connected with the antisymmetrical mode.

I. INTRODUCTION

If there is a surface with finite conductivity parallel to the plasma boundary and a surface wave is excited in the plasma, induced currents will lead to wave energy dissipation in the surface. Simple estimates show that there is a simple condition for matching "a generator and a load," when the energy absorption has a maximum. For this matching the surface conductivity σ of the absorber must satisfy the condition $\sigma = (\omega/2\pi k)$, where ω and k are respectively the frequency and the wave number of the excited wave in the plasma. The time-averaged energy flux density S to the absorbing surface is:

$$S = (\omega/k)(E^2/8\pi) \quad , \tag{1}$$

where E is the electric field amplitude in the plasma. If the wave generation is a result of a plasma instability and the increment of the instability γ is much less than the frequency, the damping introduced by the finite-conductivity surface can be sufficient to suppress the instability. The cyclotron wave in low-density plasmas with anisotropic ion velocity distribution is such an instability. The increment of this instability is approximately $\gamma \sim (m_e/m_i)^{1/3}\omega_{ci}$ (Timofeev and Pistunovich, 1967), where m_e and m_i are electron and ion mass, respectively, and ω_{ci} the ion cyclotron frequency.

The oscillation energy contained in a plasma layer with thickness of order k^{-1} increases at the rate

$$\dot{w} \sim \frac{2\gamma}{k} \frac{E^2}{8\pi} . \tag{2}$$

Comparing (1) and (2) one can see that $(\dot{w}/S) \sim (2\gamma/\omega_{ci}) \sim 2(m_e/m_i)^{1/3}$, i.e., the energy absorption rate exceeds by more than one order of magnitude the rate of increase of oscillation energy, a sufficient condition for instability suppression.

In experiments, instability suppression by means of absorptive wall encounters a number of difficulties. However, electrodes in a feedback system which provides the condition

$$\phi_e = \delta\phi_p \tag{3}$$

on the surface surrounding the plasma could work like the equivalent of the absorptive wall. Here ϕ_p is the potential of the plasma surface, ϕ_e is the electrode potential and δ is the transmission coefficient. The Poynting vector S which gives the density of the energy flux from the plasma to the surface with the condition (3) is:

$$S \sim \frac{\omega}{k} \frac{1}{2\,kd} \frac{E^2}{8\pi} |\delta| \sin\vartheta , \tag{4}$$

where d is the distance between the surface where ϕ_p is measured and the electrodes where ϕe is set up by the feedback system with the transmission coefficient δ. θ is the phase given by the feedback system. When there is a phase lag on the electrodes ($\sin\theta < 0$), the case corresponding to the surface with finite conductivity, the energy flux is directed from the plasma to the wall.

More detailed treatment of the problem based on the analysis of the dispersion equation with the condition (3) on the wall of the chamber was given by Arsenin, Zhil'tsov, Likhtenstein, and Chuyanov (1968).

II. EXPERIMENTAL CONDITIONS

The experiments were carried out in the OGRA-2 mirror device (Artemenkov, Vasilev, Galkin, Golovin, Zhil'tsov, Zubarev, Kachalov, Klochkov, Konyaev, Kuznetsov, Likhtenshtein, Maslennikov, Mukhin, Nekrasov, Panov, Pistunovich, Pustovoit, Svishchev, Semashko, Tereshkin, Tushabranishvili, Chuyanov, Chukhin, and Yushkevich, 1966) with the plasma created by fast **atom** injection

For a given condition the cyclotron instability developed in the plasma had one mode structure, without mode mixing. It was possible to go from one unstable mode structure to the other by changing the ratio ω_{oe}/ω_{ci}. Since the experiments were carried out in a simple mirror geometry, the flute instability fixed the plasma density at the level close to the cyclotron instability threshold. This allowed variation of ω_{ci} in a wide range without changing the value of ω_{oe}.

Figure 1 shows the dependence of cyclotron oscillation amplitude

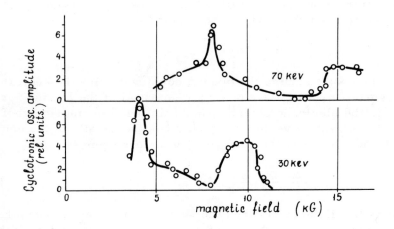

Fig. 1. Intensity of cyclotron oscillations as a function of magnetic field for two different energies of injected atoms: 70 keV and 30 keV.

on magnetic field for two fast atom injection energies - 70 and 30 keV.

From the phase measurement and the measurements of oscillation amplitude distribution along the trap axis (Fig. 2) it follows that in the first region at the highest magnetic field the antisymmetric mode oscillations are developed, and in the second region the symmetrical mode oscillations are developed. In both cases the azimuthal mode number of the wave is m = 1 and the direction of the azimuthal velocity of the wave coincides with that of the Larmor ion rotation.

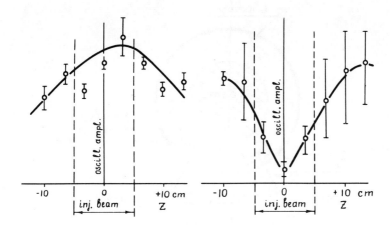

Fig. 2. Axial distributions of radial electric field intensity at cyclotron frequency for the antisymmetrical and symmetrical modes.

The displacement of oscillation amplitude maximum to the lower magnetic fields with decreased injection energy (Fig. 1) occurs because the plasma density is limited by the flute instability whose threshold is proportional to the ion energy; the same mode of oscillations sets in at the same values of ω_{oe}/ω_{ci}.

III. APPARATUS

The stabilization experiment on the symmetrical mode was carried out by means of a single-electrode feedback circuit (Fig. 3), consisting of a capacitive probe placed in the midplane of the trap, a delay line, an amplifier, and a driving electrode. The probe and the electrode placed at the radius 25 cm were also the limiters. The driving electrode had the dimensions 80 cm along the trap axis and 30° in the azimuthal direction. Thus, only ~ 10% of the plasma surface was exposed to the electrode.

For stabilization of the antisymmetrical mode a similar system was used, differing only in that the extended electrode was divided into two insulated parts, each connected to one of the symmetrical antiphase amplifier outputs, and also in that the capacitive probe was displaced from the midplane of the trap by a distance equal to one half of the plasma radius. During adjustment of the apparatus the lines delivering

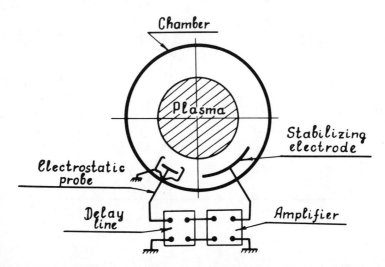

Fig. 3. Feedback circuit for stabilization of symmetrical cyclotron oscillation.

signals to the driving electrodes were balanced to give oscillations on different electrodes with equal amplitudes but opposite phases.

The cyclotron activity of the plasma was recorded by capacitive probes placed at different azimuths around the plasma. The energy spectra of the charge-exchange neutrals were measured by means of an electrostatic analyzer. Two foil detectors recorded the intensity of charge-exchange neutral flux at the wall of the chamber. One of them was placed in the midplane of the trap, the second was at a distance of 17 cm from the first. In stable operating conditions at low plasma density the side foil detector recorded a neutral flux one hundred times less than the central foil detector. Relative increase of the side detector signal gave the information on plasma expansion in the axial direction.

IV.   RESULTS

A. Symmetric Mode.

Typical oscillograms of the various processes in a feedback experiment are shown in Fig. 4 for the case of stabilization of a symmetrical mode. The upper trace shows an evelope of the flute plasma oscillations, the second trace the intensity of the injected neutral beam, the forth trace the plasma density (charge-exchange neutral flux), and the fifth trace the envelope of cyclotron oscillations.

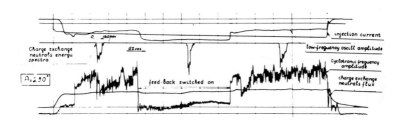

Fig. 4.   Oscillograms of diagnostic probe signals during stabilization of the symmetrical cyclotron modes.

On the third trace the energy spectra of the fast charge-exchange
atoms measured by the electrostatic analyzer are shown.   The
scanning analyzer was triggered three times per injection pulse.
The scanning duration was 0.1 sec.   In the middle of the 3-sec pulse
feedback was switched on for 1 sec.   As shown in Fig. 4 the switching
on of the stabilization circuit led to the change of the cyclotron oscillation amplitude, insignificant increase of the flute-oscillation
amplitude, very small (less than 10%) decrease of plasma density,
and appreciable narrowing of the energy spectrum of the charge-
exchange atoms.

   The increase of the flute-oscillation amplitude and the related
decrease of the plasma density are explained by the following.   The
suppression of cyclotron oscillations leads to the narrowing of the
ion spectrum and therefore to the decrease of the mean ion energy.
This, in turn, lowers the threshold of the flute instability because
$n_{cr} \sim T_{\perp i}$.   Hence the flute-oscillation amplitude increases and the
density decreases.   But the density decrease is small enough not to
interfere with measurements of the parametric dependences of the
oscillation amplitude.

The dependence of the oscillation amplitude on the amplification of feedback is shown in Fig. 5. Maximum suppression corresponds

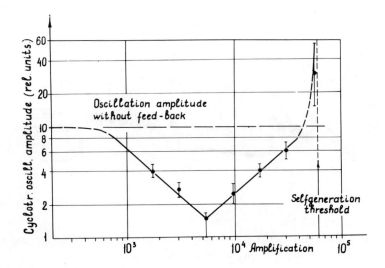

Fig. 5. Dependence of symmetrical-mode amplitude on amplification coefficient of the feedback system.

to the transmission coefficient $\delta \sim 2 - 3$ (taking into account the sensitivity of the capacitive probe). V. V. Arsenin has shown that the decrease of stabilizing effect with further increase of δ is connected to the "pushing" of the cyclotron oscillation zone into the plasma interior and which results in the decrease of energy flux from plasma to electrode. Consequently, the artificial damping decreases.

The dependence of the oscillation amplitude on phase shift for the optimal value of transmission coefficient δ is shown in Fig. 6. The oscillation amplitude is minimized when the phase angle is 300°, which corresponds to lagging of the oscillations on the electrode.

The possibility of affecting the cyclotron oscillation amplitude demonstrates the interrelation between the broadening of the energy spectrum of ions confined in the trap and the oscillation amplitude of symmetrical cyclotron modes. The spectrum width (Fig. 4) of the charge-exchange neutrals and consequently the width of the energy of ions in the trap decrease upon instability suppression. The energy

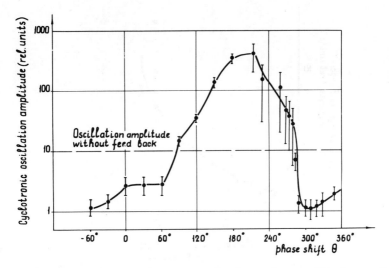

Fig. 6. Dependence of symmetrical-mode amplitude on the feedback phase shift.

spectrum shape is rather complicated and cannot be described by a halfwidth. Therefore to describe the spectrum broadening the flux intensity of neutrals with energy 140 keV (two times higher than the injection energy) was selected. The dependence of this intensity on the cyclotron-oscillation amplitude controlled by regulating the amplification of feedback is shown in Fig. 7. The decrease of the oscillation amplitude is seen accompanied by the decrease of the high-energy fraction in the energy spectrum.

B. Antisymmetric Mode.

Typical oscillograms of the probe signals for the injection pulse are shown in Fig. 8 for the antisymmetrical mode. The upper trace is the injection current, the second is the amplitude of flute oscillations, the third is the charge-exchange neutral flux measured by the central foil detector, the fourth is the signal of the side foil detector, and the fifth is the amplitude of the cyclotron oscillations. The switching on of the feedback circuit suppresses the cyclotron oscillations. Simultaneously, the signal of the side foil detector is disappearing with a time constant ~ 50 msec; the dependences of the oscillation amplitude on the phase shift and the amplification of the feedback system, shown in Figs. 9 and 10, qualitatively coincide with those found in the experiments with the symmetrical mode.

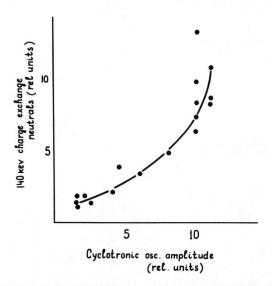

Fig. 7. Dependence of intensity of charge-exchange neutrals with energy 140 keV on the symmetrical-mode amplitude controlled by feedback.

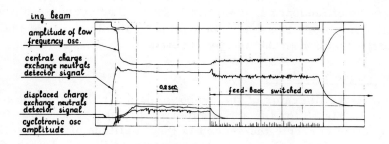

Fig. 8. Oscillogram of diagnostic probe signals during stabilization of antisymmetrical cyclotron mode.

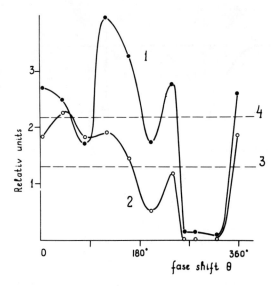

Fig. 9.   Dependences of the antisymmetrical mode amplitude (curve 1) and the displaced foil detector signal (curve 2) on feedback phase shift. Dashed lines are the levels of the cyclotron oscillations (3) and the charge-exchange neutral flux (4) when feedback system is switched off.

Maximum suppression occurs at $\delta \sim 1$ - 2. In this case the phase shift between the oscillation of the driving electrodes and the potential oscillation at the plasma surface (in front of electrodes) is approximately $-90°$. The phase interval over which stabilization takes place is $60°$. The dependences of foil detector signals on the phase shift and the amplification of the feedback indicate that the axial expansion of the plasma is a function of the oscillation amplitude of the antisymmetrical cyclotron mode.

V.   CONCLUSIONS

The described results have demonstrated the possibility of suppressing cyclotron instability in a rarefield plasma by a feedback system simulating the absorption of the oscillation energy by a surface of finite conductivity. The effectiveness of the method is underlined by the practically complete suppression of the oscillations by means of only one electrode contacting only $\sim 10\%$ of the plasma surface. By regulating the feedback parameters the resulting change

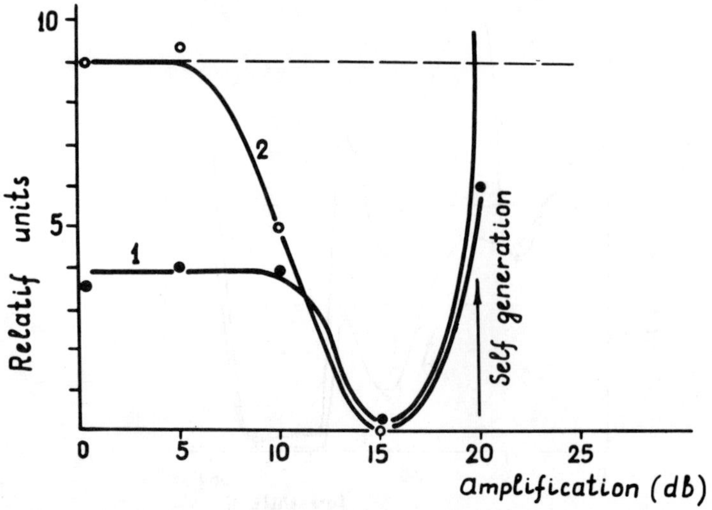

Fig. 10. Dependences of the amplitude of the antisymmetrical mode (curve 1) and side foil detector signal (curve 2)on feedback amplification.

of the cyclotron oscillation level, with other plasma parameters remaining unchanged, shows that the broadening of the energy spectrum of the ions in the trap is a result of the symmetrical mode of the cyclotron instability, and the axial expansion of the plasma is a result of the antisymmetrical mode of the instability. It also follows from these experiments that both modes are positive-energy oscillations (Timofeev and Pistunovich, 1967).

Acknowledgment

The authors consider it a pleasant duty to thank Dr. V.V. Arsenin for numerous and valuable discussions and Dr. I. N. Golovin for his interest in the work, and to express gratitude to the OGRA staff for their help during the experiments.

Part II

DYNAMIC CONTROL

DYNAMIC-STABILIZATION THEORIES FOR ELECTROSTATIC MODES

Herbert Lashinsky
Institute for Fluid Dynamics and Applied Mathematics
University of Maryland, College Park, Maryland 20742

and

Edmond M. Dewan
Air Force Cambridge Research Laboratories
Bedford, Massachusetts 01730

ABSTRACT

Methods for dynamic stabilization of plasma instabilities can be divided into two classes: 1) closed-loop and 2) open-loop. In the former case the correction forces are derived from the instability; in the latter case they are essentially independent. The two approaches are discussed and analog-computer results that demonstrate open-loop dynamic stabilization of the van der Pol oscillator are described. Possible applications to the suppression of plasma instabilities are discussed.

I. INTRODUCTION

Many of the present-day approaches to the problem of plasma stabilization make use of static magnetic fields, with stabilization being achieved through the use of spatially complicated magnetic-field configurations (shear, minimum-B, etc.). Another approach to the problem, which we will call dynamic stabilization, makes use of time-varying quantities, rather than fixed quantities. One form of dynamic stabilization makes use of the negative-feedback principle. The second approach is by applying periodic forces or by varying plasma parameters at frequencies that are not determined by the plasma instability. This approach is well-known in other fields of science and engineering, particularly in the design of control systems, where it is called asynchronous quenching or dithering, and one of the purposes of the present paper is to bring this concept to the attention of plasma physicists, with the hope that existing technology and knowledge can simulate ideas pertinent to the problem of plasma stabilization.

II. VAN DER POL MODEL

In discussing the dynamic stabilization of plasmas we shall make use of a model in which the unstable modes of an instability are represented by an ensemble of weakly coupled van der Pol oscillators (Lashinsky, 1966) since this picture provides a phenomenological description of many nonlinear phenomena that have been observed experimentally (Keen and Fletcher, 1969; Abrams, Yadlowsky, and Lashinsky, 1969). Moreover, a number of aspects of dynamic stabilization that will be discussed below derive specifically from the properties of the van der Pol oscillator.

The oscillator model is limited strictly to spatially bounded systems in which an instability gives rise to a steady-state oscillation, implying the existence of a nonlinear saturation mechanism. Each of the discrete instability modes is then treated as a weakly non-linear oscillator, which is weakly coupled to the other modes, these being treated in the same way. The central point of the analysis is the description of the time behavior of a characteristic variable x of each such oscillatory mode by the nonlinear van der Pol equation,

$$\frac{d^2x}{dt^2} - \varepsilon(\alpha - \beta x^2)\,\omega_0\,\frac{dx}{dt} + \omega_0^2 x = 0 \quad , \tag{1}$$

where ε is a smallness parameter that describes the weak non-linearity, α linear growth rate, and β nonlinear saturation parameter. The solution to Eq. (1) grows exponentially and then reaches a saturation level. This behavior stems from the nonlinear dissipation term which, in turn, derives from a nonlinear response function, such as a nonlinear plasma conductivity, provided this quantity can be expressed in a power series and provided that it exhibits saturation. The time behavior of the mode energy is then given by the solution

$$\frac{dE}{dt} = \alpha E - \beta E^2 \quad . \tag{2}$$

III. CLOSED-LOOP STABILIZATION

We divide dynamic stabilization methods into two classes: closed-loop and open-loop. Closed-loop methods are represented by the well-known negative-feedback approach: a deviation from the desired plasma configuration gives rise to an error signal which is amplified and then produces a change in some parameter in such a way as to counteract the original perturbation. However, there are three features of negative-feedback systems that should be kept in mind

in considering the possible application to plasma stabilization and
which are relevant in connection with the discussion of open-loop
methods given below. The feedback system must function as a
linear amplifier if it is to produce exact cancellation of the original
perturbation. The feedback system is a narrow-band system adjusted to provide the required phase shifts at a given instability
frequency. To suppress an instability in a plasma whose parameters
are changing in time, the parameters of the feedback must change in
time to "track" the instability. The third feature of negative-feedback appears in cases in which a plasma exhibits several simultaneously unstable modes. The quenching of one mode by feedback
can cause the growth of another because of mode competition.
Hence it is necessary that all unstable modes be quenched simultaneously.

A simple example of mode competition can be seen in a two-
mode system. When the two unstable modes are viewed as a pair of
weakly coupled van der Pol oscillators, the system is characterized
by a pair of coupled "rate" equations that describe the behavior of
the mode amplitudes A_1 and A_2:

$$dA_1/dt = a_{10}A_1 - a_{11}A_1^2 - a_{12}A_1A_2$$

$$dA_2/dt = a_{20}A_2 - a_{22}A_2^2 - a_{21}A_1A_2$$

(3)

where the a_{i0} are the linear growth rates, the a_{ii} are the damping
terms, and the a_{ij} are the mode-coupling terms. Equilibrium,
i.e., steady-state oscillation in two simultaneous modes,   corresponds to $dA_1/dt = dA_2/dt = 0$ and the equilibrium amplitude for the
j-th mode is

$$A_j^0 = \left(\frac{a_{j0}}{a_{jj}} - \frac{a_{ji}a_{i0}}{a_{ii}a_{jj}} \right) \bigg/ \left(1 - \frac{a_{ij}a_{ji}}{a_{ii}a_{jj}} \right).$$

(4)

It can be shown from Eq. (4) or from a general phase-plane
analysis that if a_{10}, the growth rate for mode 1, is reduced the
result is that A_1^0 is reduced, as expected; however, it is found that
A_2^0, the amplitude of mode 2, is increased.

IV. OPEN-LOOP STABILIZATION

The open-loop methods discussed here are nonlinear and can be applied only if the instability exhibits a saturation mechanism. The general concept is to force the system to execute oscillations that are the same as those due to the undesired instability, implying excursions into the region of nonlinear damping. However, it is forced to do so at a high frequency (compared with the instability frequency). The time average over the high-frequency excursions, as seen at the instability frequency, can then yield a net unfavorable energy balance for the instability itself. The high frequency is not determined by the instability, being arbitrary and subject only to the requirement that it be high enough so that the averaging effect is valid. We note that this feature circumvents the tracking problem discussed above in connection with the closed-loop method. Similarly, linearity is no longer important and multimode suppression is accomplished since all of the unstable modes see a net unfavorable energy balance. In this section we shall consider two open-loop methods and then consider possible ways of translating the results to the problem of plasma stabilization.

A. Asynchronous Quenching.

The model used in this prototype situation is described by the van der Pol equation with a periodic forcing term[*]:

$$\frac{d^2x}{dt^2} - \varepsilon(\alpha - \beta x^2)\,\omega_0\,\frac{dx}{dt} + \omega_0^2 x = E \sin \Omega t \tag{5}$$

with the solution of the form

$$x \simeq A \sin \Omega t + B \sin \omega_0 t \quad ; \quad A \simeq \frac{1}{1-(\Omega-\omega_0)^2} \,,\, B \simeq (1-A^2)^{1/2}. \tag{6}$$

Here B is a measure of the oscillation amplitude at the instability frequency, determined by the value of A, which is a measure of the response of the system at the forcing frequency. Since we are considering situations in which $\Omega \gg \omega_0$, the resonance denominator in A is a large quantity and A itself is small. For this reason it is difficult to make A reach the value unity (in the normalization used here) so that it is difficult to quench the oscillation at the instability frequency.

[*] E. M. Dewan and H. Lashinsky, to be submitted for publication.

In any case, even if the instability is quenched, the system executes
oscillations at the forcing frequency; these oscillations may be
averaged out on the slow time scale of the instability and in the
plasma case, would hopefully be less troublesome from the point of
view of diffusion and plasma loss.

The situation described here is known in the literature as
asynchronous quenching and one of the standard treatments by
Minorsky (1962) did not include the result that the system is left
oscillating at the frequency of the forcing function, making the
situation appear overly optimistic. It was later pointed out that
asynchronous quenching is actually what is known as frequency entrainment at frequencies far from resonance. (Pengilley and
Milner, 1967; Dewan and Lashinsky, 1969) The fact that the system
is left oscillating at the high frequency has been verified by analog-
computer experiments (Dewan and Lashinsky, 1969).

B. Dithering

This method is different from asynchronous quenching in that
an external forcing term is not applied. Rather, the system state
is caused to vary periodically by varying a parameter that is not
associated with the resonance properties. Details of the dithering
concept are given in a recent text on control engineering (Gelb and
Vander Velde, 1968). As a nonplasma example, in the case of a
vacuum-tube oscillator the grid bias would be varied periodically at
a frequency much higher than the oscillation frequency, thus driving
the system into the nonlinear damping region of the vacuum-tube
characteristic. The prototype equation is the modified van der Pol
equation

$$\frac{d^2x}{dt^2} - \epsilon\{\alpha - \beta(x + E \sin \Omega t)^2\} \{\omega_0 \frac{dx}{dt} + \Omega E \cos \Omega t\} + \omega_0^2 x = 0 . \qquad (7)$$

The solution is of the form

$$x \simeq A \sin \Omega t + B \sin \omega_0 t \; ; \; A \simeq \frac{1}{1-(\Omega-\omega_0)^2} , B \simeq (1-E^2)^{1/2} . \qquad (8)$$

The difference between asynchronous quenching and dithering
lies in the coefficient B in Eq. (6) and Eq. (8). Suppression of
the instability corresponds to B → 0. In asynchronous quenching
[Eq. (6)] it is difficult to make B → 0, as was pointed out above. In
dithering [Eq. (8)], on the other hand, it is only necessary to satisfy
the condition E → 1, by simply increasing the excursion of the parameter being varied, without the need to overcome the effect of a

resonance denominator.

Analog-computer experiments have been carried out to verify the dithering concept as applied to the van der Pol oscillator and it is found that the oscillator can indeed be quenched in this way. In these experiments the analog computer is set up in accordance with Eq. (7). A typical result is shown in Fig. 1, in which the ratio of the quenching frequency to oscillator frequency is approximately 5 to 1.

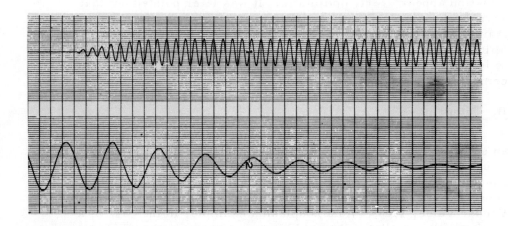

Fig. 1. Analog-computer result showing damping of the van der Pol oscillator by the application of a dithering signal. Time increases from left to right. The lower trace shows the oscillator signal while the upper trace shows the gradual application of the dithering signal. In this example the ratio of dithering frequency to oscillator frequency is approximately 5 to 1 and the oscillator is quenched in about 7 oscillator cycles.

V. PLASMA STABILIZATION

It has been shown in this work that the van der Pol oscillator can be damped by the application of a dithering signal. On the other hand, as has been indicated above, a number of plasma instabilities in bounded plasmas can be represented phenomenologically in terms of the van der Pol oscillator. The next logical step would be to set up the appropriate analogous physical conditions in a plasma to repro- duce the conditions for which damping is achieved in the analog- computer experiments. In order to accomplish this purpose, however,

it is necessary to have a concrete picture of the details of the instability mechanism so that the elements in the plasma that are analogous to the oscillator can be identified. Merely knowing that a certain
plasma dispersion equation has unstable roots for certain values of
the parameters does not provide the required information. As an
example of what is needed we note that there are a number of plasma
instabilities that arise by virtue of the transfer of energy of directed
longitudinal motion of electrons that move in synchronism with a
wave characteristic of the collective motion of the plasma viz., the
two-stream instability, collisionless drift instability, etc. One might
then conceive of coupling to the electrons in such a way as to dither
the velocity of the synchronous electrons in order to make them
extract energy from the wave as well as to provide energy to it, and
to make the net average energy exchange unfavorable for excitation
of the instability. Thus the instability would be quenched although
the electrons would be executing oscillatory motions about the mean
longitudinal motion.

5.2 DYNAMIC STABILIZATION OF A CONFINED PLASMA

A. Samain

Association Euratom-CEA, Centre d'Etudes Nucleaires
Fontenay-aux-Roses (France)

ABSTRACT

The stabilization of a magnetically confined plasma by the presence of a H.F. potential is discussed. The amplitude of the H.F. potential may be reduced if it is modulated with a short period along the direction of the density gradient.

I. INTRODUCTION

Among the dynamic stabilization schemes of a magnetically confined plasma reported, the simplest is probably the action of the so called H. F. potential induced by the oscillating field: under certain conditions, the slow components of the particle motion may be calculated by adding a fictitious hamiltonian H_m independent of time to the hamiltonian H_s governing the particle motion in the static field (Artsimovich, 1964). Eventually H_m is attracted towards the center of the plasma, and is not very sensitive to slow perturbations of the plasma equilibrium. Then the plasma may be thermodynamically stable (Demirkhanov, 1968). It can be shown that drift modes are stabilized if H_m increases across the plasma by a quantity δH_m of the same order as its temperature T. Flute stabilization is obtained with $\delta H_m \sim T(\Delta/R)$ where Δ^{-1} is the logarithmic derivative $\partial n/n\, \partial x$ of the plasma density and R^{-1} is the curvature of the magnetic lines of force.

We present a variant of this scheme. Assume that H_m is not equivalent to a smooth potential well for the plasma particles as above, but rather is modulated along the x direction of the density gradient with a short period π/k_m much smaller than Δ. This is the case if the oscillating field has a standing wave structure with the period $2\pi/k_m$ along x. The plasma, while confined at scale length Δ, is assumed in thermodynamic equilibrium in each of the small-scale potential wells associated with the x modulation of H_m. This structure requires a certain amount of energy to be smoothed down. Therefore, it tends to stabilize the drift modes and the flute modes localized along ox in an interval $\gg k_m^{-1}$, which would result in such smoothing. Actually stabilization of these modes is obtained for a smaller amplitude of the oscillating field than in the preceding scheme: H_m needs to be of the order of $T/k_m \Delta$ and $T/k_m (\Delta R)^{1/2}$, respectively, instead of T and $T(\Delta/R)$.

II. EQUILIBRIUM

As a specific model, we consider a low-β collisionless plasma slab confined along x by a uniform magnetic field B parallel to z. The magnetic curvature R^{-1} is simulated by a gravitational force G giving drift velocities v_c along oy to the particles:

$$v_c = \Delta R^{-1} v_d \ll v_d \quad ; \tag{1}$$

$v_d = c T (\Delta q B)^{-1}$. The oscillating field is a small-amplitude electric field E_m parallel to x, with a standing wave structure along ox:

$$E_m(x,t) = \frac{\partial \psi_m(x,t)}{\partial x} = E_m \sin(k_m x) \exp(i\omega_m t) + c.c. + O(E_m^2). \tag{2}$$

The field E_m is an eigenmode of oscillation specified in particular by the boundary conditions $\psi_m(x,t) = 0$ in two planes x = const. enclosing the plasma slab and by the local dispersion relation $\mathcal{R}(\omega_m, k_m) = 0$. We assume that $\omega_m \tilde{\omega} \omega_{ci}$ and $k_m \rho_{thi} \sim 1$, where $\omega_c = (-qB/m c)$, $\rho_{th}^2 = 2T/m\omega_c^2$, and the index i refers to ions. We assume further that the field E_m is not resonant with the particles. Then, the reported formalism (Rebut and Samain, 1969a; Rebut and Samain, 1969b Samain, 1969) is applicable. We describe each particle in the presence of the field E_m by action variables ρ_c, $v_{\parallel}$, and x_c, functions of the coordinates and the conjugate momenta of the particle and of the time t. They may be given the following interpretation: if the amplitude of E_m is adiabatically reduced to 0, a particle initially described by the action variables ρ_c, $v_{\parallel}$, x_c ends in the resulting field B with a Larmor radius. equal to ρ_c, with a velocity along z equal to $v_{\parallel}$, and with a guiding center abscissa equal to x_c. The variables ρ_c, $v_{\parallel}$, and x_c are constants of motion and specify the average properties of the particle motion. In particular, the time-averaged cyclotronic frequency Ω_c and the time-averaged velocities v_y and v_z of a particle described by ρ_c, $v_{\parallel}$, and x_c are given by:

$$\Omega_c(\rho_c, v_{\parallel}, x_c) = \frac{2}{m\omega_c} \frac{\partial H}{\partial \rho_c^2} \quad , \tag{3}$$

$$v_z(\rho_c, v_{\parallel}, x_c) = \frac{1}{m} \frac{\partial H}{\partial v_{\parallel}} \quad , \tag{4}$$

$$v_y(\rho_c, v_\parallel, x_c) = - \frac{1}{m\,\omega_c} \frac{\partial H}{\partial x_c} \quad , \tag{5}$$

where $H(\rho_c, v_\parallel, x_c)$ is a function of $\rho_c, v_\parallel$, and x_c specific of the particle assembly. This function is the fictitious hamiltonian $H_s + H_m$ mentioned above. Discarding the static electric field of the order of E_m^2 for simplicity we have

$$H(\rho_c, v_\parallel, x_c) = \frac{1}{2} m\,\omega_c^2 \rho_c^2 + \frac{1}{2} m v_\parallel^2 + \frac{q\,v_c B}{c} x_c + H_m \quad , \tag{6}$$

$$H_m(\rho_c, x_c) = \frac{E_m^2}{k_m^2} \left\{ \sum_\ell \ell \, \frac{cq}{B} \, \frac{d}{d\rho_c^2} \, [J_\ell^2(k_m \rho_c)] \frac{(-1)^\ell}{\omega_m + \ell\omega_c} \right\} \cos(2k_m x_c). \tag{7}$$

These are valid as long as $\xi_m = cE_m(B\omega_m)^{-1} \ll k_m^{-1}$. Note that H_m is modulated with respect to x with the period π/k_m.

Each assembly of particles of the plasma at equilibrium is described by a distribution function $F(\rho_c, v_\parallel, x_c)$: $F\,d\rho_c^2\,dv_\parallel\,dx_c$ is the number of particles in the domain $d\rho_c^2\,dv_\parallel\,dx_c$. We assume that for each assembly F has the form: $F = A \exp(-H/T + x_c/\Delta)$ where A and T are constants. These distribution functions effectively describe a confined plasma with a temperature T and an average $\partial n/n\partial x \approx \Delta^{-1}$. They may be shown to be produced by collisions between particles.

III. STABILITY

An electrostatic mode perturbing this equilibrium state is described by a potential of the form:

$$\psi(x, y, z, t) = \exp(i\omega t + ik_y y + ik_z z) \sum_n \psi_n(x) \exp(in\omega_m t) + c.c. \tag{8}$$

We restrict ourselves to low frequency ($\omega \ll \omega_{ci}$) modes with wave number $k_x = [- d^2\psi_o(x)/dx^2]^{1/2} \ll k_m$. Assume that $k_m \lambda_D$, where λ_D is the plasma Debye length, is much smaller than 1 throughout the plasma. Then the dispersion relation $\mathcal{R}(\omega_m, k_m)=0$ of the oscillating field E_m is independent of the plasma density, so that perturbations of the latter have a small influence on the structure E_m. For these conditions it is possible to show that if $k_m \xi_m \ll 1$, the terms $n \neq 0$ in the Eq. (8) play a negligible role in the dispersion relation giving the frequency ω of the mode. This dispersion relation may be written as:

$$\sum_{\substack{\text{ions} \\ \text{electrons}}} \iiint F\left(\rho_c, v_{\parallel}, x_c\right)\left\{1 - \frac{\omega + k_y v_d}{\omega + k_y v_y + k_z v_z} [J_0(k_\perp \rho_c)]^2\right\} d\rho_c^2 \, dv_{\parallel} \, dx_c = 0 \qquad (9)$$

where $k_\perp^2 = k_x^2 + k_y^2$ and v_y and v_z are functions of ρ_c and x_c specified by Eqs. (4) - (7).

In the presence of the oscillating field the function v_{yi} depends on ρ_c and x_c instead of being the constant v_{ci}. This prevents Eq. (9) from having unstable solutions ω. This happens for all the modes under consideration when the dispersion of the values taken by $v_y(\rho_c, x_c)$ is of the order of v_{di}. If $k_m \rho_{thi} < 1$, the necessary value of E_m is given by:

$$m_i \frac{E_m^2 c^2}{B^2} \approx \frac{T}{k_m \Delta} \left| \frac{\omega_m - \omega_{ci}}{\omega_{ci}} \right| .$$

The stabilization of the M.H.D. flute modes, for which $|\omega| \gg k_y v_{gi}$, needs only $m_i (E_m^2 c^2/B^2) > \left\{4T/[(\Delta R)^{1/2} k_m]\right\} \cdot |\omega_m/\omega_{ci} - 1|$. The corresponding values of H_m are $\sim T/k_m \Delta$ and $T/k_m (\Delta R)^{1/2}$, respectively. The power per particle to maintain the field E_m in the plasma against the action of the collisions is of the order of νH_m, where ν is the ion collision frequency.

5.3 DYNAMIC STABILIZATION OF DRIFT WAVES IN A COLLISIONLESS PLASMA BY A.C. ELECTRIC FIELDS

M. Dobrowolny
Laboratori Gas Ionizzati, EURATOM, CNEN, Frascati, Italy

ABSTRACT

The influence of an a.c. electric field parallel to the confining magnetic field on low-frequency instabilities in an inhomogenous collisionless low-β plasma has been considered. Stabilization effects have been found for a majority of unstable drift waves. Corresponding stability criteria in terms of the electric field amplitude and conditions on the modulation frequency are given.

I. INTRODUCTION

A number of authors(Fainberg and Shapiro, 1967; Ivanov, Rudakov and Teichmann, 1968) have examined dynamic stabilization of drift-type instabilities by means of high frequency fields. More recently, (Dobrowolny, Engelmann and Levine, 1969a, 1969b) some attempts have been made to consider also possibilities of dynamic stabilization by means of low amplitude, low frequency fields (the modulation frequency being comparable to the mode frequency to be stabilized, and much smaller than the ion cyclotron frequency).

In the present work the effect on low-frequency instability of an a.c. electric field parallel to magnetic field lines in investigated for a collisionless low-β plasma. The physical model considered is that of an inhomogeneous plasma (both density and temperature gradients are allowed) confined by a uniform magnetic field. An alternating electric field $E_0(t) = E_0 \cos \Omega_m t$, uniform in space and parallel to the magnetic field is present; correspondingly, there are alternating particle currents.

II. THEORY

A linear stability theory of such an equilibrium is developed with respect to electrostatic perturbations with frequency smaller than the ion cyclotron frequency. The theory is done for small amplitudes of the dynamic modulation (in the sense that maximum particle velocities in the ac field have to be smaller than corresponding thermal velocities). The modulation frequency is assumed smaller than the ion cyclotron frequency. The problem of stability is solved as an initial value problem starting from the drift kinetic equation (for ions and electrons) and using Laplace transformation in time for the perturbed quantities. In this way the non-linear interaction between the a. c. field and the plasma modes, in particular the generation of beat frequencies $\omega \pm n\Omega_m$ (ω denoting a plasma eigenfrequency and n being an integer) is correctly described.

The derived dispersion relation, including the effect of the modulation on the equilibrium, can be schematically written as

$$\epsilon\,(\omega) + \alpha^2\,F(\omega) = 0. \tag{1}$$

Here $\alpha^2 = u_e^2/v_e^2$ where u_e is the maximum electron velocity in the a. c. field and v_e is the electron thermal velocity. $\epsilon\,(\omega) = 0$ is the dispersion relation valid for time-independent equilibria, $\alpha^2 F(\omega)$ the correction due to the modulation. (The explicit expression for $F(\omega)$, in terms of integrals over particle parallel velocities, is complicated and will be reported elsewhere.) The effect of the modulated equilibrium on the dispersion law is thus second order in α which simply means that a double beating is needed to modify the dynamics of any considered mode: a first beating between the mode and the modulation generates satellites with frequencies $\omega \pm \Omega_m$, which in turn beat again with the modulation to produce a contribution to the original mode.

III. APPLICATIONS

Drift instabilities driven by resonant electrons in the regime $v_i < |\omega/k_z| < v_e$ (k_z being the parallel wave number and $v_j = (2KT_j/m_j)^{1/2}$) can be stabilized (at $\alpha \ll 1$) at modulation frequencies $\Omega_m > k_z v_e$. In the case of the "universal" instability, the stability criterion can be written

$$\frac{\alpha^2}{s\left(\dfrac{T_e}{T_i} - \dfrac{1}{2}\dfrac{k_z^2 v_e^2}{\Omega_m^2}\right)}\left[\frac{1}{2}\frac{k_z^2 v_e^2}{\omega_{en}^*\Omega_m}\left(s + \frac{1}{2}\frac{k_z^2 v_e^2}{\Omega_m^2}\right) + \frac{k_z^2 v_e^2}{\Omega_m^2} - 2s\frac{T_e}{T_i}\right] \geq s\left(1 + \frac{T_e}{T_i}\right) \tag{2}$$

with the additional condition on the frequency of modulation

$$\frac{1}{2} \frac{k_z^2 v_e^2}{\Omega_m^2} \left(\frac{\Omega_m}{\omega_{en}^*} s + 2 + \frac{1}{2} \frac{k_z^2 v_e^2}{\omega_{en}^* \Omega_m} \right) > 2 s \frac{T_e}{T_i} \tag{3}$$

Here $s = 1/2 k_\perp^2 a_i^2$, $a_i = (2 K T_i m_i / e^2 B^2)^{1/2}$ is the ion Larmor radius, and $\omega_{jn}^* = k_y (K T_j / m_j \Omega_{cj}) (n_0'/n_0)$ is the drift frequency related to the density gradient. The analogous stability conditions in the case of the instability due to an average electron current u and/or a temperature gradient (when $d\ell n T / d\ell n n < 0$) are

$$\alpha^2 \left[\frac{\Omega_m}{\omega_{en}^*} + \frac{2\Omega_m^2}{k_z^2 v_e^2} (1 - |\xi|) \right] > |\xi| \quad , \tag{4}$$

and

$$\frac{\Omega_m}{\omega_{en}^*} + \frac{2\Omega_m^2}{k_z^2 v_e^2} > \frac{2\Omega_m^2}{k_z^2 v_e^2} |\xi| \tag{5}$$

where $\xi \equiv (\omega_{eT}^* + 2k_z u)/\omega_{en}^*$ and ω_{jT}^* is the drift frequency related to temperature gradients.

In the case of the drift instability due to a large temperature gradient ($d\ell n T / d\ell n n \gg 1$), in the same frequency regime $v_i < |\omega/k_z| < v_e$, it has been found that, at modulation frequencies $\Omega_m \sim \omega \ll \omega_{jT}^*$ the unstable root which one has in the absence of the a.c. field ($\mathrm{Im}\,\omega \sim \mathrm{Re}\,\omega \sim (k_z v_i)^{2/3} \omega_{eT}^* {}^{1/3}$), is eliminated for field amplitudes such that α^2 is very close to $k_z^2 v_e^2 / \omega_{eT}^* {}^2$ (which requires temperature gradients such that $|\omega_{eT}^*/k_z v_e| \gg 1$).

Finally the drift instability which one has in the regime $\omega/k_z > v_e$ ($\omega \sim \pm i \sqrt{2} |k_z v_e|/\sqrt{s}|$ at $k_z^2 v_e^2/\omega_{en}^* {}^2 s \ll 1$), is found to be stabilized at modulation frequencies $\Omega_m > \omega$, provided the modulation amplitude and frequency satisfy the following conditions:

$$\alpha^2 \geq 2\sqrt{2}\, s^{3/2} \left| \frac{\Omega_m}{k_z v_e} \right| \left(1 + \frac{\omega_{eT}^*}{\omega_{en}^*} \right)^2 , \quad 1 < \left| \frac{\Omega_m}{k_z v_e} \right| < s^{-3/2} \tag{6}$$

Contrary to the drift wave cases mentioned above, no effect of the time modulation of the equilibrium is found on purely transverse modes ($k_z = 0$) at any modulation frequency so that, in particular, no effect has to be expected on unstable flute modes in the presence of a curvature of the magnetic lines.

5.4 DYNAMIC STABILIZATION OF THE DRIFT DISSIPATIVE INSTABILITY BY AN INHOMOGENEOUS R.F. FIELD

D. Lépéchinsky, P. Rolland, and J. Teichmann
CEN/Saclay S.I.G. France

ABSTRACT

The effect of an R.F. electric field of frequency Ω oriented parallel to a static magnetic field, and having a gradient in the radial direction of the plasma column, on the drift dissipative instability is studied using two-fluid equations. When Ω is greater than the collision frequency of the electrons, the presence of a gradient of the R.F. field reduces the minimum value of this field required for complete stabilization by more than an order of magnitude when compared with the value for the case of zero R. F. gradient.

I. INTRODUCTION

The new method proposed by Lépéchinsky and Rolland (1969) for the treatment of the problem of stabilization of the drift dissipative instability is based on a macroscopic two-fluid model of the plasma with scalar pressure, constant electron temperature T_e, cold ions, and electron collision frequency ν_e ($\nu_i = 0$). Stabilization may be achieved by use of a uniform RF electric field E_{RF}, parallel to the static magnetic field H_o directed along the plasma column. This method has been developed because of certain difculties and lack of clarity found in the work of Fainberg and Shapiro (1967) on the same subject. The method considers the nonlinear interaction of low-frequency (ω) drift waves with an RF field of frequency Ω. This interaction is represented by quadratic terms which are neglected by Fainberg and Shapiro such as $\vec{V}_{RF} \cdot \nabla n^e$, $(\vec{V}_{RF} \cdot \nabla) \vec{v}$, where $\vec{V}_{RF}$ is the RF velocity of the electrons and $\vec{v}$ and n^e are the velocity and density fluctuations, respectively. These terms give rise to fluctuations at frequencies of $\omega \pm \Omega$. Rearranging terms in the equations so that all of them have the same time dependence, one obtains a new dispersion relation containing an RF term which is stabilizing in a broad frequency band under appropriate conditions. In the present work, the method is extended to the case where the applied RF field, still parallel to $\vec{H}_o$, has a gradient in the radial direction.

II.   MODEL USED AND GOVERNING EQUATIONS

The reference system is that shown in Fig. 1 where $\vec{H}_o$ and $\tilde{E}_{RF}$ are parallel to the z axis. The $\tilde{E}_{RF}$ and density gradients are parallel to the x axis and the RF magnetic field $\tilde{H}_{RF}$ is along the y axis.

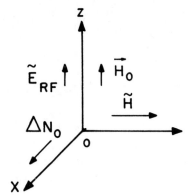

Fig. 1

The fields $\tilde{E}_{RF}$ and $\tilde{H}_{RF}$ have the form:

$$\tilde{E}_{RF} = \bar{E}_z (x) \sin \Omega t \tag{1a}$$

$$\tilde{H}_{RF} = \bar{H}_y (x) \cos \Omega t. \tag{1b}$$

The Maxwell relation $\nabla \times \vec{E} = -1/C\, (\partial \vec{H}/\partial t)$ gives :

$$\tilde{H}_{RF} = - \frac{C}{\Omega} \frac{\partial \bar{E}_z}{\partial x} \cos \Omega t \tag{2a}$$

and $\quad \bar{H}/\vec{H}_o = - \dfrac{e E_z}{m\Omega} \dfrac{\kappa_E}{\omega_{ce}}$ $\hspace{2cm}$ (2b)

where

$$\kappa_E = \frac{1}{E_z} \frac{\partial E_z}{\partial x} \quad \text{and} \quad \omega_{ce} = \left| \frac{e H_o}{m C} \right| .$$

We assume that

$$\Omega \ll \omega_{ce}, \quad \omega < \nu_e \ll \omega_{ce}, \quad \text{and} \quad \bar{H}/H_o \ll 1. \tag{3}$$

The two-fluid macroscopic equations used arè as follows:

$$\frac{\partial N_\alpha}{\partial t} + \nabla \cdot (N_\alpha \vec{V}_\alpha) = 0 \tag{4}$$

$$N_\alpha \left\{ \frac{\partial \vec{V}_\alpha}{\partial t} + (\vec{V}_\alpha \cdot \nabla) \vec{V}_\alpha - \frac{e_\alpha \vec{E}}{m_\alpha} - \frac{e_\alpha}{m_\alpha c} \vec{V}_\alpha \times \vec{\mathcal{H}} + \nu_\alpha V_\alpha \right\} + \frac{T_\alpha}{m_\alpha} \nabla N_\alpha = 0 \tag{5}$$

where

$$\alpha = e, i, \quad e_\alpha = \mp e, \quad \vec{E} = \vec{E}_o + \tilde{\vec{E}}_{RF}, \quad \text{and} \quad \vec{\mathcal{H}} = \vec{H}_o + \tilde{\vec{H}}_{RF}.$$

The plasma will be assumed neutral and with no D.C. electric field ($E_o = 0$). Setting $N_e = N_o + N_{RF}$ and $\vec{V}_e = \vec{V}_D + \vec{V}_{RF}$ and solving the zero order equations of motion and continuity, one obtains:

$$N_{RF} = 0; \quad V_D^y = - \frac{T_\kappa}{m\omega_{ce}}; \quad V_D^x = \frac{\nu}{\omega_{ce}} V_D^y; \quad V_D^z = 0. \tag{6}$$

If $\Omega \gg \nu$ then $V_{RF}^z \cong \nu_z^{RF} \cos \Omega t$; $V_{RF}^y \cong v_z^{RF}(\bar{H}/H_o) \cos^2 \Omega t$;

$$V_{RF}^x \cong - (\Omega/\omega_{ce}) (\bar{H}/H_o) v_z^{RF} \sin 2 \Omega t \tag{7}$$

where

$$v_z^{RF} = |e\bar{E}_z / m\Omega| \quad \text{and} \quad \kappa = (1/N_o)(\partial N_o / \partial x).$$

The expression for V_{RF}^y shows that this component of the RF velocity has always a non-zero mean value averaged over the RF period, which is proportional to the RF "pressure" :

$$\left\langle V_{RF}^y \right\rangle = \frac{1}{2} v_z^{RF} \frac{\bar{H}}{H_o} = - \frac{1}{\omega_{ce}} \frac{e^2}{4m^2 \Omega^2} \frac{\partial}{\partial x} (\bar{E}_z^e).$$

This term opposes the natural drift velocity V_D^y when the RF electric field and density gradients are of opposite sign. We shall see that this condition must be satisfied for stabilization.

Using the "adiabatic" approximation   (Lépéchinsky and Rolland, 1969) the first-order equation of motion of the electrons is:

$$\frac{\partial \vec{v}}{\partial t} + \underline{(\vec{V_T} \cdot \nabla) \vec{v}} + (\vec{v} \cdot \nabla) \vec{V_T} - \frac{e\nabla\phi}{m} + \vec{v} \times \omega_{ce}(1 + \frac{\tilde{H}}{H_o})$$

$$+ \nu_e \vec{v} + \frac{T}{m} \frac{\nabla n^e}{N_o} = 0 \tag{8}$$

where $V_T = \vec{V_D} + \vec{V}_{RF}$ and the electric field first-order perturbation is $E_1 = -\nabla\phi$.

The first order equation of continuity reads:

$$\frac{\partial n^e}{\partial t} + N_o \nabla \cdot \vec{v} + n^e \underline{\vec{\nabla} \cdot \vec{V}_D} + n^e \nabla \cdot \vec{V}_{RF} + \vec{V}_{RF} \cdot \nabla n^e + \vec{V}_D \cdot \nabla n^e + \vec{v} \cdot \nabla N_o$$

$$= 0 \tag{9}$$

It can be shown that the underlined terms can be neglected. We now expand the perturbation terms n^e, $\vec{v}$ and ϕ into a series:

$$n^e, \vec{v}, \phi = \sum_{n=-\infty}^{+\infty} n_n^e, \vec{v}_n, \phi_n \exp(-i\omega_n t + i\vec{k} \cdot \vec{r}) \tag{10}$$

where

$$\omega_n = \omega + n\Omega \quad \text{and} \quad n = 0, \pm1, \pm2 \ldots$$

Substituting into Eqs. (8) and (9), one sees that terms such as $n^e \nabla \cdot \vec{V}_{RF} = -n^e V_{RF}^x \kappa$ or $\vec{V}_{RF} \cdot \nabla n^e = i K n^e \cdot \vec{V}_{RF}$, where the $\vec{V}_{RF}$ components are given by Eq. (7), introduce indices $n \pm 1$ and/or $n \pm 2$, if every term of the equations has to have an $[\exp(-i\omega_n t)]$ time dependence. Thus the term $n^e \cos \Omega t = (n^e/2) \cdot (e^{i\Omega t} + e^{-i\Omega t})$ gives the contribution $(1/2)(n_{n+1}^e + n_{n-1}^e)$.

Assuming $k_x = 0$ (i.e. neglecting radial oscillations) one finally obtains the following equation for the electrons:

$$n_n^e \left\{ - i\omega_n + \frac{Tk_z^2}{m(\nu - i\omega_n)} \right\} - N_o \frac{e\phi_n}{m} \left\{ \frac{k_z^2}{\nu - i\omega_n} \right.$$

$$+ \frac{ik_y \kappa \omega_{ce} \left[1 - i\frac{k_z}{\kappa} \frac{\omega_{ce} \tilde{H}/H_o}{\nu - i\omega_n} \right]}{\omega_{ce}^2 + (\nu - i\omega_n)^2} \left. \right\} + n_n^e ik_y \frac{\overline{H}}{H_o} \frac{v_z^{RF}}{2}$$

$$+ ik_z \frac{v_z^{RF}}{2} (n_{n+1}^e + n_{n-1}^e) = 0 \qquad (11)$$

A similar equation, though of much simpler form is obtained for the ion density fluctuation n_n^i. It has been assumed in this latter derivation that $T_i = \nu_i = 0$ and that the RF field has no influence on the ions.

Using quasi-neutrality assumption $n_n^e = n_n^i$ and eliminating ϕ_n from Eq. (11) and the corresponding equation for the ions, the following dispersion relation is obtained:

$$n_n A_n - \alpha_n \left[k_y \frac{\overline{H}}{H_o} n_n + k_z B_n (n_{n+1} + n_{n-1}) \right] = 0 \qquad (12)$$

where A_n, B_n and α_n are functions of ω_n and other plasma parameters.

The infinite set of equations represented by (12) is then solved first by limiting (12) to three equations only (with $n = 0, \pm 1$) and then neglecting n_{+2} and n_{-2} ($n_{\pm 2} \ll n_o$). This gives the final dispersion relation:

$$\omega^2 + i\omega (Dk_z^2 + \omega_s) - i\omega_s \omega_* - \frac{v_{RF}^2 k_y k_E}{2\omega_{ce}} (\omega_o - \omega) - \frac{v_{RF}^2 k_z^2}{2} (\omega_o - \omega)$$

$$\left(1 + \frac{\Omega^2}{\nu_e^2}\right) \left\{ \frac{\left(\alpha + \frac{\omega_o}{\Omega^2} \xi_{HF} \right) \left[\omega_o + \omega_{oE} \left(1 - \frac{\Omega^2}{\omega_{ce}^2} \right) \right] + i\omega_s \beta - \xi_{HF}}{\Omega^2 \left(\alpha + \frac{\omega_o}{\Omega^2} \xi_{HF} \right)^2 - \left(i\omega_s \beta - \xi_{HF} \right)^2} \right\} = 0. \qquad (13)$$

where:

$$\alpha = 1 + \frac{\Omega^2}{\nu_e^2} - \frac{\omega_s}{\nu_e}\left(1 - \frac{\Omega^2}{\omega_{ce}^2}\right) \; ; \; \beta = 1 - \frac{\Omega^2}{\omega_{ci}^2} + i\frac{\omega_e \Omega^2}{\omega_s \omega_{ci}^2}\left(1 + \frac{\Omega^2}{\nu_e^2}\right) \; ;$$

$$\xi_{HF} = -\frac{v_{HF}^2}{2} k_y \frac{\kappa_E}{\omega_{ce}}\left(1 + \frac{\Omega^2}{\nu_e^2}\right) \; ; \; \omega_o = -\frac{\kappa \omega_{ci}}{k_y} \; ;$$

$$\omega_{oE} = \frac{\kappa_E \omega_{ci}}{k_y} \; ; \; \omega_s = \frac{k_z^2}{k_y^2}\frac{\omega_{ce}\omega_{ci}}{\nu_e} \; ; \; D = \frac{T}{m\nu_a}$$

In the absence of the RF electric field gradient $(\kappa_E = 0)$ Eq. (13) reduces to the usual dispersion relation (Lépéchinsky and Rolland, 1969). With no RF present Eq. (13) becomes the well known dispersion relation of the drift dissipative instability (Kadomtsev, 1965).

III. STABILITY CRITERIA

A. High RF Frequencies $(\Omega > \omega_{ci}, \nu_e)$.

For stability:

$$\frac{\omega_*^2}{\omega_{ci}^2} \leq \frac{v_{RF}^2}{2}\left[-\frac{\kappa\,\kappa_E}{\omega_{ce}\omega_{ci}} + \frac{k_z^2}{\Omega^2}\left(1 - \frac{\kappa_E}{\kappa}\right)\right], \tag{14}$$

where $\omega_* = k_y V_D^y$. Here the following inequalities have been assumed: $\omega_* < \omega_o$, ω_{oE}; $\omega_s < \omega_{ci}$; and $\nu_e \sim \omega_{ci}$. The RF field gradient has a dominant stabilizing effect provided that κ and κ_E are of opposite sign. The minimum required RF field intensity E_{RF} (min) corresponding to Eq. (14) is found to be 20 times less than in the case of a uniform field (Lépéchinsky and Rolland, 1969), i.e., of the order of 1 volt/cm with the usual plasma parameters.

B. Intermediate RF Frequencies $(\omega_{ci} > \Omega > \nu_e)$.

For stability:

$$\frac{\omega_*^2}{\omega_{ci}^2} \leq \frac{v_{RF}^2}{2}\left[-\frac{\kappa\,\kappa_E}{\omega_{ce}\omega_{ci}} + \frac{k_z^2}{\Omega^2}\frac{\kappa^2}{k_y^2}\left(1 - \frac{\kappa_E}{\kappa}\right)\right], \tag{15}$$

when ν_e, $\omega_s < \omega_{ci}$ and $\nu_e \sim \omega_s$. This applies to the case of high magnetic fields. With $\Omega \sim \omega_{ci}/2$ one finds $E_{RF(min)}$ to be of the order of 0.5 Volt/cm.

C.   Low RF Frequencies ($\Omega < \nu_e$).

Here the effect of the electric field gradient is greatly reduced and one obtains a result similar to that found previously (Lépéchinsky and Rolland, 1969). It should be noted that a kinetic treatment rather than the macroscopic approach used here made by one of us (P.R.) leads to conclusions in agreement with those indicated above except in case(c) when $\Omega < \nu_e$.

IV.   CONCLUSION

We conclude with the following remarks. First, it has been found that the present stabilization method reduces anomalous diffusion by a factor $k_z^2 / k_\perp^2$ (i.e. by more than 10^{-4}) and second   the stabilizing RF field together with its gradient can be produced by means of current carrying conductors placed outside of the plasma column, parallel to its axis.

5.5 INFLUENCE OF HIGH-FREQUENCY FIELD
ON PLASMA INSTABILITIES

A. A. Ivanov and V. F. Muraviev
Kurchatov Institute of Atomic Energy
Moscow, USSR

ABSTRACT

The results of theoretical investigations of high frequency stabilization of plasma instabilities are reported. The ordinary wave and helicon-type wave are shown to suppress electrostatic instabilities with frequencies small compared to the frequency of the stabilizing wave. The comparison of two modes of stabilization has shown that for "loss-cone" instability helicon stabilization is more preferable. An explanation of the stabilization mechanism is given.

I. INTRODUCTION

It is well known that when plasma is immersed in a strong magnetic field H_{oz} the most dangerous instabilities are electrostatic ones with $k_z \gg k_\perp$. If the charged particles of one kind oscillate with frequency Ω large compared to that of the instability ω in the direction perpendicular to the magnetic field, the potential perturbations will be reduced and the instability will be suppressed. In an ordinary wave and in a helicon-type wave such motion of charged particles is observed. In the first case thermal movement of particles along the instantaneous direction of magnetic field lines eliminates the perturbation potential. In the second case such an elimination occurs due to the drift motion of electrons in the transverse direction.

II. THEORY

Let's find the perturbation of the electron distribution function for the plasma inhomogeneous in the X direction, immersed in the magnetic field $\vec{H} = \{ 0, H_1 \cos \Omega t, H_o \}$, $H_1 \ll H_o$, $\Omega \ll \omega_{ce}$. The drift kinetic equation is used and the motion of electrons along z due to electric field E_{1z} is neglected. The perturbations of potential ϕ and distribution function f_1 will have the form

$$\phi(\vec{r},t) = \sum_{n=-\infty}^{+\infty} \phi_n \exp i\,(\vec{k}\cdot\vec{r} - \omega t - n\Omega t)$$

$$f_1(\vec{r},t) = \sum_{s=-\infty}^{+\infty} f_s \exp i\,(\vec{k}\cdot\vec{r} - \omega t - s\Omega t)$$

(1)

Thus we have

$$f_s(v_{\parallel}) = -\frac{e}{m}\frac{1}{v_{\parallel}}\frac{\partial f_o(v_{\parallel})}{\partial v_{\parallel}} +$$

(2)

$$\sum_{p=-\infty}^{+\infty} \phi_p \frac{J_p(\mu v_{\parallel})\,J_s(\mu v_{\parallel})}{k_z v_{\parallel} - \omega} \left[\frac{ck_y}{H_o}\frac{\partial f_o}{\partial x} - \frac{e}{m}\frac{1}{v_{\parallel}}\frac{\partial f_o}{\partial v_{\parallel}}(\omega + p\Omega)\right]$$

Here $J_p(\mu v_{\parallel})$ is the Bessel function, $\mu = k_y H_1/\Omega H_o$ and $v_{\parallel}$ is the velocity of electrons along the instantaneous direction of the magnetic field line. Formula (2) is valid for $|\omega - k_z V_{Te}| \ll \Omega$.

If the helicon wave propagates along z with frequency Ω ($\omega_{ci} \ll \Omega \ll \omega_{ce}$) and wave number $k_o = [\omega_{pe}/c][(\Omega/\omega_{ce})^{1/2}]$, the field components will be:

$$\vec{H}_1(z,t) = H_1\left[\vec{e}_x \sin(k_o z - \Omega t) + \vec{e}_y \cos(k_o z - \Omega t)\right]$$

(3)

$$\vec{E}_1(z,t) = \frac{\Omega}{k_o c} H_1\left[\vec{e}_x \cos(k_o z - \Omega t) - \vec{e}_y \sin(k_o z - \Omega t)\right]$$

In this case the perturbations of potential and distribution function are:

$$\phi(\vec{r},t) = \sum_{n=-\infty}^{+\infty} \phi_n \exp i\{-\omega t + \vec{k}\cdot\vec{r} + n(k_o z - \Omega t)\}$$

(4)

$$f_1(\vec{r},t) = \sum_{s=-\infty}^{+\infty} f_s \exp i\{-\omega t + \vec{k}\vec{r} + s(k_o z - \Omega t)\}$$

For $k_\perp^2 v_{Te}^2 / \omega_{ce}^2 \ll 1$ the solution of the kinetic equation will be as follows:

$$f_s(v_z) = \frac{e\phi_s}{m} \frac{k_\perp^2}{\omega_{ce}^2} f_o(v_z) + \frac{e\phi_s}{T} \frac{v_z f_o(v_z)}{v_z - \Omega/k_o}$$

$$+ \frac{e}{T} \left[\frac{v_z(\omega k_o - k_z \Omega)f_o(v_z)}{(k_z v_z - \omega)(k_o v_z - \Omega)} + \frac{ck_y T}{eH} \frac{1}{k_z v_z - \omega} \frac{\partial f_o(v_z)}{\partial x} \right]$$

$$\sum_{n=-\infty}^{+\infty}(-1)^{n+s} \; \phi_n \, J_n(\nu) \, J_s(\nu) \, e^{i(n-s)\beta} \tag{5}$$

Here

$$\nu = \frac{k_\perp H_1}{k_0 H_0} \; ; \quad \beta = \text{arctg} \left(-\frac{k_x}{k_y} \right) , \; |k_z v_{Te} - \omega| \ll |k_o v_{Te} - \Omega| .$$

III. APPLICATION

A. Loss-Cone Instability (Rosenbluth and Post, 1965)

Substituting from Eq. (5) the perturbation of electron density into the Poisson equation, and taking into account that the helicon wave does not influence the motion of ions, we obtain the infinite set of equations for ϕ_n. Because of the property of the Bessel functions $\sum_{s=-\infty}^{+\infty} J_s^2 = 1$, this system reduces to the following dispersion relation:

$$\left[1 + \frac{\omega_{pe}^2}{\omega_{ce}^2} - \frac{\omega_{pe}^2 k_o^2}{\Omega^2 k^2} - \frac{\omega_{pi}^2}{k^2 v_{Ti}^2} \, F\left(\frac{\omega}{kv_{Ti}}\right) \right]$$

$$\left(1 + \frac{\omega_{pe}^2}{\omega_{ce}^2} - \frac{\omega_{pe}^2}{\omega^2} \frac{k_z^2}{k^2} \right)$$

$$+ J_o^2(\nu) \frac{\omega_{pi}^2}{k^2 v_{Ti}^2} \, F\left(\frac{\omega}{ku_{Ti}}\right) \frac{\omega_{pe}^2}{k^2} \left(\frac{k_o^2}{\Omega^2} - \frac{k_z^2}{\omega^2} \right) = 0 \tag{6}$$

For Eq. (6) to be valid the following conditions must be satisfied: Ω/k_o, $\omega/k_z \gg v_{Te}$. It follows from Eq. (6), that when $H_1/H_o \lesssim y\,(v_{Ti}/c)$, ($\omega_{ci}/\omega_{pi} = 0,2$), the condition for instability ($\gamma > \omega_{ci}$) fails to be satisfied. It should be noted that the condition for stabilization, obtained by Ivanov, Rudakov, and Teichmann (1968), in which an ordinary wave was used, is $(H_1/H_o) > (v_{Ti}/v_{Te})$.

B. Loss-Cone Drift Instability (Mikhailovsky, 1966)

The following dispersion relation is obtained in a similar way:

$$\left(1 + \frac{\omega_{pe}^2}{\omega_{ce}^2} + \frac{1}{k^2\lambda_{di}^2}\,\frac{\omega^*}{\omega}\right)\left(1 + \frac{\omega_{pe}^2}{\omega_{ce}^2} - \frac{\omega_{pe}^2 k_o^2}{\Omega^2 k^2}\right)$$

$$= i\,\frac{\omega}{|k|\,v_c}\,\frac{b}{k^2\lambda_{di}^2}\left\{1 + \frac{\omega_{pe}^2}{\omega_{ce}^2} - \frac{\omega_{pe}^2 k_o^2}{\Omega^2 k^2} + \left[1 - J_o^2(\nu)\right]\left(\frac{\omega_{pe}^2 k_o^2}{\Omega^2 k^2} + \frac{\omega^*}{\omega}\,\frac{1}{k^2\lambda_{di}^2}\right)\right\},$$

(7)

where $v_c \sim v_T$, $b \underset{\sim}{2} 1$. It follows from Eq. (7) that the growth rate γ is now $\gamma = J_o^2(\nu)\,\gamma_c$, where γ_o is the growth rate of instability in the absence of the R. F. field.

C. Universal Instability (Rudakov and Sagdeev, 1961)

Using Eqs. (1), (2) and the condition of quasi-neutrality one obtains the infinite set of equations:

$$2\phi_s + \frac{i\sqrt{\pi}}{k_z v_{Te}} \sum_{p=-\infty}^{+\infty} \phi_p\left[\omega + \omega_e^*\left(1 - \frac{\eta_e}{2}\right)\right].$$

$$J_p\left(\mu\,\frac{\omega}{k_z}\right) J_s\left(\mu\,\frac{\omega}{k_z}\right) - \frac{\omega + \omega_i^*}{\omega}\,A_i\phi_s\,\delta_{so} = 0,$$

(8)

where

$$A_i = \int_{-\infty}^{+\infty} J_o^2(\mu v_{\parallel})\,f_{vi}(v_{\parallel})\,dv_{\parallel}.$$

The following assumptions were made: $T_i \sim T_e$, $\eta_i \equiv d\ell n T_i/d\ell n \eta = 0$, $\Omega \gg k_z v_{Te}$, $\mu v_{Ti} \ll 1$. We have taken into account that $\omega \ll \Omega$, i.e. $\phi_{-s} = (-1)^s \phi_s$, hence $\sum_{p=-\infty}^{+\infty} p\phi_p J_p = 0$. In this case we obtain the dispersion relation:

$$2\omega - (\omega + \omega_i^*) A_i + \frac{1\sqrt{\pi}}{k_z v_{Te}} \left[\omega + \omega_e^* (1 - \frac{\eta_e}{2}) \right]$$

$$\times \left\{ \omega J_o^2 (\mu \frac{\omega}{k_z}) + \frac{1}{2} \left[1 - J_o^2 (\mu \frac{\omega}{k_z}) \right] \left[2\omega - (\omega + \omega_i^*) A_i \right] \right\} = 0$$

(9)

The dominant contribution is seen to be defined by the zero harmonic of expansion (1) (the first term in the braces), because $\omega \ll k_z v_{Te}$. In fact , we have the same result as that by Mikhailovsky (1966):

$$\omega = - \frac{\omega_e^* A_i}{2 - A_i}; \quad \gamma = \frac{\sqrt{\pi}}{|k_z| v_{Te}} \frac{A_i \omega_e^{*2}}{(2-A_i)^3} J_o^2 (\mu \frac{\omega}{k_z}) \left[2(1 - A_i) + \frac{\eta_e}{2} (A_i - 2) \right]$$

(10)

D.   Drift Temperature Instability (Rudakov and Sagdeev, 1961)

Assuming the electrons to be distributed according to $n_s/n_o = e\phi_s/T$ and obtaining the perturbation of the ion density from Eq. (2), we again have an infinite set of equations

$$\left(1 + \frac{T_i}{T_e}\right) \phi_s - \sum_{p=-\infty}^{+\infty} \phi_p \int_{-\infty}^{+\infty} \frac{\omega_i^* \left(1 - \eta/2 + \eta \frac{v_{||}^2}{v_{Ti}^2}\right) + \omega}{\omega - k_2 v_{||}}$$

$$J_p (\mu v_{||}) J_s (\mu v_{||}) f_{oi} (v_{||}) dv_{||} = 0.$$

(11)

We cannot reduce the system to one equation. However, when $\mu v_{Ti} \gg 1$, the infinite determinant converges rapidly. In this case, omitting all terms in (1) except $n = 0$ corresponds to the zero approximation (Mikhailovsky, 1966).

5.6 ELECTRON-ION COLLISION EFFECTS ON PARAMETRIC INSTABILITY[*]

R. A. Stern

Bell Telephone Laboratories, Whippany, New Jersey 07981

and

Takaya Kawabe[†] and P. K. Kaw

Plasma Physics Laboratory, Princeton University
Princeton, New Jersey 08540

ABSTRACT

The choice of high frequency fields in dynamic stabilization schemes is often limited, because the stabilizing field may serve as a pump for driving parametric instabilities in the plasma. We investigate the dependence of the parametric instability growth rate on electron-ion collisions in a plasma. Sharp enhancement of the complex plasma conductivity near $\omega \approx \omega_p$, in the presence of ion correlations, is shown to have dramatic effects on this growth rate.

I. INTRODUCTION

Calculations of the high-frequency conductivity σ in fully ionized plasmas including electron-ion encounters have revealed that, near $\omega \sim \omega_{pe}$, both the real and the imaginary components, σ_R and σ_I, of σ exhibit resonances associated with the generation of longitudinal plasma oscillations (Dawson and Oberman, 1962, 1963). We are concerned with some large effects which can result from this behavior. It will be shown that, in the presence of ion correlations, the wave energy may become negative near ω_{pe}. Such waves absorb energy from the system. Their excitation by deliberately introducing ion correlation may provide a useful

---

[*] This work is partially supported by Atomic Energy Commission Contract AT(30-1)-1238.

[†] On leave of absence from Institute of Plasma Physics, Nagoya University, Nagoya, Japan.

stabilization technique. In particular, decay (parametric) instability growth rates, which depend strongly on the derivatives of σ_I, will be appreciably affected by electron collisions when one of the modes involved occurs near ω_{pe}.

II.   THEORY

Consider a simple idealized case of an infinite, homogeneous electron-ion plasma supporting two finite but small-amplitude longitudinal oscillations $\underset{\sim}{E}_1(\omega_1)$, $\underset{\sim}{E}_2(\omega_2)$. A strong "stabilizing" longitudinal field $\underset{\sim}{E}_o(\omega_o)$ is applied. Coupling between these three fields is expressed in the parametric approximation by the well-known set of equations:

$$(i\omega_1 \epsilon_o + \sigma_1)\underset{\sim}{E}_1 + \underset{\sim}{E}_o \underset{\sim}{\chi}_1 \underset{\sim}{E}_2 = 0$$

$$(i\omega_2 \epsilon_o + \sigma_2)\underset{\sim}{E}_2 + \underset{\sim}{E}_o \underset{\sim}{\chi}_2 \underset{\sim}{E}_1 = 0 \tag{1}$$

where $\underset{\sim}{\chi}_1$ and $\underset{\sim}{\chi}_2$ are the susceptibilities. The condition for existence of finite fields is the vanishing of the determinant of the coefficients in Eq. (1).

The usual procedure is to expand about the resonant frequencies [†]

$$\sigma_{1,2} \approx \sigma_{1,2\,R,I}\left(\omega_{1,2}^o\right) + \left.\frac{\partial\sigma_{1,2R}}{\partial\omega}\right|_{\omega=\omega_{1,2}^o} \cdot \Delta\omega$$

$$+ i \left.\frac{\partial\sigma_{1,2I}}{\partial\omega}\right|_{\omega=\omega_{1,2}^o} \cdot \Delta\omega \tag{2}$$

Introducing the preceding into the determinant, it is found that $\Delta\omega$ obeys a quadratic equation with the solution (here $\partial\sigma_R/\epsilon_o d\omega$ is assumed to be much less than 1, and is neglected):

[†] The subscripts R, I denote real and imaginary parts of the complex conductivity.

$$\Delta \omega = \frac{i}{2 \epsilon_o} \frac{1}{1 + \dfrac{\partial \sigma_{1I}}{\epsilon_o \partial \omega}} \frac{1}{1 + \dfrac{\partial \sigma_{2I}}{\epsilon_o \partial \omega}}$$

$$\times \left[\left[\sigma_{1R} \left(1 + \frac{\partial \sigma_{2I}}{\epsilon_o \partial \omega} \right) + \sigma_{2R} \left(1 + \frac{\partial \sigma_{1I}}{\epsilon_o \partial \omega} \right) \right] \right.$$

$$\pm \left\{ -4E_o^2 \chi_1 \chi_2 \left(1 + \frac{\partial \sigma_{1I}}{\epsilon_o \partial \omega} \right) \left(1 + \frac{\partial \delta_{2I}}{\epsilon_o \partial \omega} \right) \right.$$

$$\left. \left. + \left[\sigma_{1R} \left(1 + \frac{\partial \sigma_{2I}}{\epsilon_o \partial \omega} \right) - \sigma_{2R} \left(1 + \frac{\partial \sigma_{1I}}{\epsilon_o \partial \omega} \right) \right]^2 \right\}^{1/2} \right] \tag{3}$$

The expansions are valid only if $\Delta \omega \ll \omega$, and if higher derivative of σ_I are small. We will be principally interested in the case when σ_I has a cusp-like behavior near $\omega = \omega_{pe}$. At the peak, the expansion is strictly not correct. On either side of the peak however the solution may still hold. We see then from Eq. (3) that the sign of $\Delta \omega$ is strongly dependent on the factors $1 + \partial \sigma_I / \epsilon_o \partial \omega$ in the denominator. These represent, in the near-equilibrium case, the wave energies. The magnitude of the effect we consider here can therefore be intuitively seen: if $\partial \sigma_{1I} / \epsilon_o \partial \omega$ can take on the value -1, while the other factor as well as the numerator do not change signs, the value of $\Delta \omega$ can change from positive to negative. This expresses the fact that, in the presence of negative energy waves, the application of the strong field E_o may lead to parametric suppression.

We now estimate conditions under which the derivatives may be negative and sufficiently large. Dawson and Oberman have computed the high-frequency impedance $(R + iX)$ of a fully ionized plasma regarding the ions as a set of stationary discrete scattering centers. Using their results, we find that, since $\sigma_I \cong -1/X$ ($X \gg R$ for most plasmas)

$$\therefore \frac{\partial \sigma_I}{\epsilon_o \partial \omega} = \frac{1}{\epsilon_o X^2} \frac{\partial X}{\partial \omega} \cong 1 + I$$

where

$$I = \frac{2Ze^2}{3\pi\omega} \int_0^{k_{max}} dk\, k^2\, \frac{\partial}{\partial\omega} \left[Re\, \frac{1}{D(k,\omega)} \right] \left\langle |n_i(k)|^2 \right\rangle$$

Here $\langle \ \rangle$ denotes the effect of ion correlations. For randomly distributed ion, $\langle \ \rangle$ = 1. For strongly nonthermal ion spectra, e.g. an ion wave spectrum peaked at $k \approx \bar{k}$ and with a width $\Delta k \approx \bar{k}$, we find:

$$I \cong I_{random}\, \frac{\delta^2}{4\pi}\, (n_i \bar{\lambda}^{-3})$$

where $\bar{\lambda} = (2\pi/\bar{k})$ and δ^2 is the normalized mean square of the ion density fluctuations. Using the numerical values of Dawson and Oberman for randomly distributed ions one finds:

$$I_{random} \cong -10^{-2}\, \frac{1}{n_e \lambda_D^3}$$

We note that in the absence of ion correlations $|I_{random}|$ can never exceed 2 and hence the denominator of Eq. (3) can never be negative. This is understandable since in this case there is no source of free energy and one cannot possibly excite negative energy waves.

On the other hand, in the presence of strong ion correlations, we get:

$$|I| \cong 10^{-3}\delta^2 \left(\frac{\bar{\lambda}}{\lambda_D} \right)^3$$

Thus if $\bar{\lambda} \gg \lambda_D$, i.e. when ion waves giving rise to ion correlations have wavelengths large compared to the Debye length, $|I| > 2$ even for comparatively weak ion density fluctuations (5 - 10%), and the effect described above can take place.

III. CONCLUSION

Whenever dynamic stabilization schemes involve the application of strong high-frequency electric fields, parametric decay processes can be important. We have shown that sufficiently strong ion correlations can alter these decay processes considerably; in particular, the normally parametrically unstable waves may be parametrically suppressed. These considerations may, therefore, be of use in the choice of high-frequency fields for some dynamic stabilization schemes.

DYNAMIC-STABILIZATION THEORIES FOR MHD MODES

6.1 DYNAMIC STABILIZATION OF HIGH-β PLASMAS

F. Troyon
Centre de Recherches en Physique des Plasmas
Lausanne, Switzerland

ABSTRACT

A discussion of various types of dynamic stabilization schemes for high-β plasmas with sharp boundary is presented. The influence of the plasma model on stability is illustrated by exact numerical calculations which indicate the problems of parametric excitation of resonances. As a particular case the results of Wolf's fluid experiments are reproduced.

I. INTRODUCTION

There exist three basic confinement schemes for a "thin-skin" $\beta = 1$ plasma, based upon the principle of "dynamic stability" (Fig. 1), (Weibel, 1960; Haas and Wesson, 1967; Berge, 1970; Orlinski, Osovets, and Sinitsyn, 1965). In the longitudinal scheme the oscillating component of the magnetic field causes the plasma surface to oscillate at the same frequency.

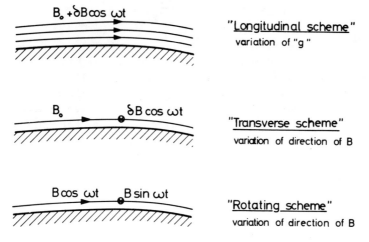

Fig. 1. The three basic schemes.

For our purpose we simulate this motion and the eventual change in curvature by an equivalent acceleration, normal to the plasma sur-face "g". In the second and third schemes the change of direction of the magnetic field is the stabilizing agent. Surface-wave charac-teristics are indeed strongly anisotropic with respect to the direction of the magnetic field. In all three cases the idea is the same; oscil-late "g" or the direction of B in such a way that all possible modes of deformation are stable at least during a fraction of the time. A given mode of deformation (either plane or cylindrical geometry) satisfies an equation of the form (Weibel, 1960 Troyon, 1967)

$$\int_{o}^{t} R(t-t') \, \dot{y}(t') \, dt' \; + \; (X + A\cos\omega t)y(t) = o \; ; \quad A > o \qquad (1)$$

In the third scheme ω should be replaced by 2ω. $R(t)$ depends only on the plasma properties and X, A on magnetic configuration. The fre-quency ω comes into the equation through the term $\cos \omega t$. This equation is strictly correct for the third scheme and approximately correct (small oscillating field) in the other schemes. It could be generalized further by adding harmonics of the frequency ω. The first scheme is the hardest to fit into this general frame, although in the fluid case (treated below) this equation is completely correct.

Denoting the Laplace transform of $R(t)$ by $\tilde{R}(s)$, $\tilde{R}(jy)$ is simply related to the acoustic impedance at the frequency y by

$$\tilde{R}(jy) = Z(y) = \frac{-\delta p(y)}{u(y)} \, ,$$

where $\delta p(y)$ is the amplitude of the pressure variation at frequency y and $u(y)$ the amplitude of the surface velocity caused by it. Know-ledge of $Z(y)$ is sufficient to determine the stability of Eq. (1) (Troyon, 1967). If one considers all possible modes the following statements can be made: X is unbounded on the positive side and bounded from below by a function depending on "g" and the magnetic field configuration. A has the same properties, but A-X is bounded from above.

II. STABILITY ANALYSIS

A. Limiting Cases

We consider two limiting cases: a) $R(t) = Z\delta(t)$, $\tilde{R}(s) = Z$, Z being a constant. This is the value obtained with the "bounce model" for a semi-infinite plasma. Equation (1) reduces to a first order equation and the stability condition is $X \geqslant 0$. This result is independent of A and ω. b) $R(t) = Z\dot{\delta}(t)$, $\tilde{R}(s) = Zs$. This is the value obtained with an inviscid incompressible semi-infinite fluid. Equation (1) reduces to a Mathieu equation. Since $\Delta = A-X$ is bounded while A and X are not, we draw all the stability diagrams in the plane ω, X for fixed values of Δ. Figure 2 shows qualitatively how the stability region of the Mathieu equation evolves as Δ increases. For a given Δ, only the region $X > -\Delta$ exists.

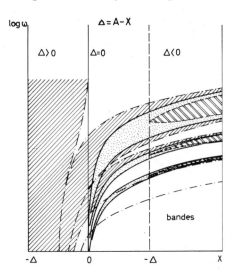

Fig. 2. Stability diagrams for the Mathieu equation $\ddot{y} + (X + (X + \Delta) \cos \omega t) y = 0$. Qualitative dependence on $\Delta = A - X$. Only 3 bands are shown. For each Δ only the region $X > -\Delta$ has a meaning. Three cases are shown with $\Delta < 0$. $\Delta = 0$ and $\Delta > 0$.

Only three bands are shown. There are an infinite number of them below. They become narrower and closer as ω decreases. The behaviour for $\Delta < 0$, $A < X$ is easily interpreted in terms of parametric excitation of the resonance $\omega = (X/Z)^{1/2}$. The interesting domain is $\Delta > 0$, $A > X$. As Δ increases the unstable regions in $X > 0$ keep growing and the bands almost merge. For $\omega \to \infty$, $X \geqslant 0$ is the stability condition, which justifies calling this condition the condition of "average stability" (Weibel, 1960) . For $X < 0$ the only sizeable stable region is limited by a line asymptotic to $X = 0$ and by the first band. This line can be obtained by a "quasi-potential" solution of the Mathieu equation (Wolf, 1970a, 1970b).

B. Influence of Viscosity. Interpretation of Wolf's Fluid
Experiments

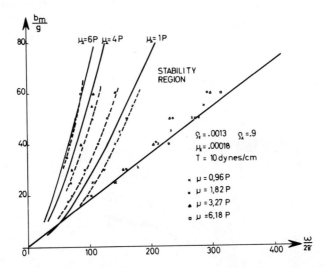

Fig. 3a. Plot of Wolf's experimental results, compared with
curves calculated for an equivalent square tube. ρ and μ designate
respectively the density and the viscosity of the two fluids and T
the surface tension of the heavy fluid.

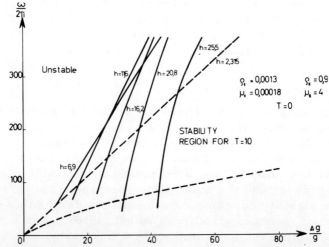

Fig. 3b. Graph illustrating the progressive deformation of
the "quasi-potential" stability boundary at high wave numbers
h for T = 0.

The introduction of viscosity into the incompressible fluid model narrows the unstable bands and suppresses them at sufficiently large values of X. The first band always extends farthest. For X < 0, viscosity lowers the boundary of the first band and pushes back towards X = 0 the other boundary. In the limit of very large viscosity one recovers the results obtained with the bounce model. This shows that Wolf's experiments (1970a, 1970b) cannot be explained by viscosity alone. Surface tension is another important factor. Figure (3a) shows Wolf's experimental points, compared with calculated values, for a surface tension T = 10 dynes/cm. In these numerical results no attempt has yet been made to find the best values of T. Figure (3b) shows how the domain of stability for T=10 disappears for T=0, due to short wavelength instabilities.

C. Collisionless Plasma

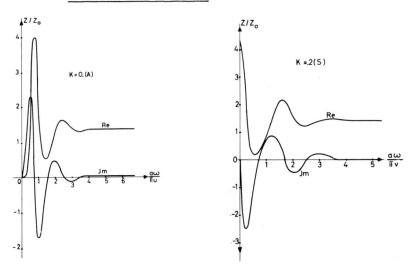

Fig. 4a,b. Plot of real and imaginary parts of the acoustic impedance $Z(\omega) = R(j\omega)$ of a collisionless plasma slab of hydrogen of thickness 2a. $Z_o = \rho u$, $u = (2kT/m)^{1/2}$, where ρ is the density and m the ion mass.
a) antisymmetric mode of deformation with k = 0.
b) symmetric mode of deformation with ak/π = .2 .

A plasma is not an incompressible fluid. In many cases of interest the plasma is better described by a Vlasov equation. We

have computed [†] $Z(y) = \tilde{R}$ (iy) for an isothermal plasma slab of
hydrogen with thickness 2a, using a variety of wavenumbers and
starting from the Vlasov equation. There are two classes of
modes corresponding to symmetric (S) and anti-symmetric (A)
deformations of the two surfaces of the slab. Fig. (4a,b) show
$Z(y)$ for $k = 0$ (A, plane analogue to $m = 1$, $k = 0$ in a cylinder)
and for $k = .2\pi/a$ (S, plane analogue to $m = 0$, same k). In Fig.
(4a) $Z(y)$ behaves as expected: as an incompressible fluid at low
frequency and as the bounce model at high frequency, with two
damped resonances visible near the average transit frequency and
at the first harmonic. In Fig. (4b) $Z(y)$ behaves differently at low
frequency. The first peak in the imaginary part has changed sign;
when $k \rightarrow 0$ this peak eventually becomes a pole, thus reflecting the
adiabatic equation of state.

Using the method described previously (Troyon, 1967) the
stability diagrams for these two cases can be computed for various
values of Δ (Fig. 5a,c). The main band structure in Fig. (5a)
looks inded like the Mathieu band structures, with dissipation, but
there is a striking difference. There is an additional band which
corresponds to the second resonance and there is no stability region for $X < 0$. This is certainly due to the second resonance.
Figure (5b) shows that, as k is increased a domain of stability
appears for $X < 0$, as the second resonance progressively disappears. For large k the bounce model result is recovered.

Figure (5c) shows the stability diagrams for various values
of Δ in the symmetric case. There is again a band structure
ascribed to two resonances, but with different characteristics.
This is not surprising since the plasma does not behave at all like
an incompressible fluid.

Thus, even for the $m = 1$ modes, the incompressible model,
correct in predicting some parametric excitations, can lead to
wrong conclusions. The real plasma has more "structure" and thus
more resonances can be excited parametrically.

D. Comparison of the Various Schemes:
The example shown allow some comparisons between the
three schemes. The schemes differ only through the values of X
and A.

[†]F. Troyon, to be published.

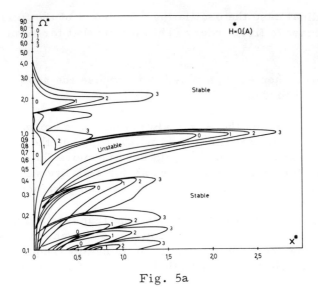

Fig. 5a

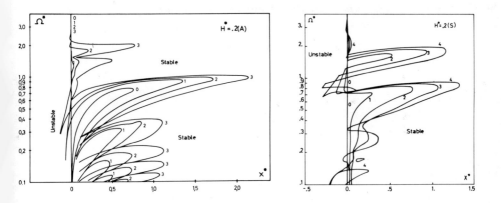

Fig. 5b   Fig. 5c

Stability diagrams obtained using the acoustic impedance of a collisionless plasma slab of hydrogen of thickness 2a. The values of X, ω, A and k are normalized:

$$\Omega^* = \frac{a\omega}{\pi u} \ , \quad X^* = \frac{aX}{\pi p_o} \ , \quad \Delta^* = \frac{a\Delta}{\pi p_o} \ , \quad H^* = \frac{ak}{\pi} \ .$$

p_o is the equilibrium plasma pressure.
a, b) The curves 0,1,2,3 correspond to $\Delta^* = 0$, 0.25, 0.5, 1 respectively. c) The curves 0,1,2,3,4 correspond to $\Delta^* = 0$, 0.25, 0.5, 0.75, 1 respectively.

238   F. TROYON

In the longitudinal case the oscillating component of the field acts only on A, at least in first order. This means that for some modes at least $X < 0$.

In the other two schemes it is possible to choose the geometry such that $X > 0$ for all modes.

In the first two schemes there will always be some amount of field inside the plasma. If $\beta \approx 1$ this field will not change the conclusions. If $\beta < 1$ the plasma behaviour should become more fluid-like (but not incompressible) which is not favorable for stability since it will lead to additional resonances which can be excited parametrically.

From the point of view of stability alone the transverse scheme is the least demanding on geometry and frequency although the results for the rotating scheme show the possibility of using any frequency if certain geometrical conditions are satisfied.[†]

ACKNOWLEDGMENT

This work was supported by the Swiss National Foundation at the Centre de Recherches en Physique des Plasmas.

[†]F. Troyon, to be published.

6.2 ON DYNAMIC STABILIZATION OF MHD INSTABILITIES

H. Wobig and H. Tasso
Institute fur Plasmaphysik GmbH, 8046 Garching, Germany

ABSTRACT

The stability of the periodic solutions of the time-dependent MHD equations in the Eulerian form is investigated. The Floquet theory for systems with periodic coefficients is generalized to this problem. An expansion procedure with respect to the amplitude ε of the oscillating solution about a static equilibrium is applied to calculate the stability conditions.

I. INTRODUCTION

In a recent paper on dynamic stabilization Berge (1969) considered the ideal MHD equations in the Lagragian formulation, by deriving an integral principle similar to the energy principle of magnetohydrodynamics. In the following paper we start from the MHD-equations in Eulerian formulation and proceed analogously to the method of normal modes.

The basic problem of the theory of dynamic stabilization is to solve a linear equation with periodic time-dependent coefficients. If it is assumed that the dynamic equilibrium is oscillating around a static equilibrium, the amplitude of this oscillation being ε, our solution can be expanded with respect to ε. This expansion is analogous to the theory of Mathieu's equation (Meixner and Schafke, 1954). Another mathematical problem is the generalization of Floquet's theory to hyperbolic partial differential equations. This will be done in Section II. In Section III the stability conditions for dynamic equilibrium are derived.

II. GENERAL THEORY

We write the nonlinear equations in the following form

$$N(S,p)\frac{d\underline{v}}{dt} = -\underline{\nabla}p + \underline{j} \times \underline{B}$$

$$\frac{d\underline{B}}{dt} = (\underline{B} \cdot \nabla)\underline{v} - \underline{B}\,\text{div}\,\underline{v}$$

$$\frac{1}{\gamma p}\frac{dp}{dt} = -\,\text{div}\,\underline{v} \tag{1}$$

$$\frac{dS}{dt} = 0$$

Here $S = pN^{-\gamma}$, $\underline{v}$ = velocity, $\underline{B}$ = magnetic field, p = pressure, N = density, $\underline{j} = \nabla \times \underline{B}$, div $\underline{B}$ = 0, and γ = adiabatic constant. This system is of the form

$$A_o(f) \frac{df}{dt} = \sum_{i=1}^{3} A_i(f) \frac{\partial f}{\partial x_i} , \tag{2}$$

$f = (\underline{v}, \underline{B}, p, S)$. The 8-by-8 matrices are symmetric matrices. If $F_0(x, t)$ is a periodic solution of (2) (dynamic equilibrium) with period T, the linearized perturbation equations of (2), which determine the stability of $F_0(x, t)$, are of the type (after some transformation)

$$\frac{\partial g}{\partial t} = \sum_{i=1}^{3} V_i(x,t) \frac{\partial g}{\partial x_i} + Cg , \tag{3}$$

where V_i and C are time-dependent periodic matrices and V_i is symmetric. System (3) is a linear, symmetric hyperbolic equation; and there always exists an unique solution to the Cauchy initial value problem. The solution of Eq. (3) can be written as:

$$g(t) = Q(t) e^{Bt} g(0). \tag{4}$$

Q(t) is a periodic (period T) and B a constant operator. Stability of the dynamic equilibrium is determined by the eigenvalues of operator B, which are called characteristic exponents. Proof of relation (4) starts from the solution operator Y(t) ($g(t) = Y(t) g(0)$), which always exists according to the standard existence theorem of hyperbolic equations. If matrices $V_i(x, t)$ and $C(x, t)$ are continuous functions, we find from Eq. (3)

$$\frac{\partial}{\partial t} \| g \|^2 \geq - \alpha \| g \|^2 , \qquad \alpha < \infty \tag{5}$$

$\| g \|^2 = \int g^* g d^3 x$. This guarantees that $Y^{-1}(t)$ exists. Because of the periodicity of $V_i(t)$ and C (t) we find from Eq. (3)

$$Y(t + T) = Y(t) Y(\tau) \tag{6}$$

We define BT = $\ell n Y(T)$ and $Q(t) = Y(t)e^{-Bt}$. From Eq. (6) we find $Q(t + T) = Q(t)$ and therefore $Y(t) = Q(t) e^{Bt}$, which completes the proof of Eq. (4).

If the dynamic equilibrium $F_0(x,t)$ is that oscillating at amplitude ε in the neighbourhood of a static equilibrium, we can expand

$$H(t) = \sum_{i=1}^{3} V_i \frac{\partial}{\partial x_i} + C \text{ in a series of } \varepsilon :$$

$$H(t) = H_o + \varepsilon H_1(t) + \varepsilon^2 H_2(t) + \ldots \tag{7}$$

the equation $\frac{\partial Y}{\partial t} = H(t) Y$ can be solved by perturbation methods:

$$Y(T) = e^{H_o T} \left[1 + \varepsilon \int_0^T \overline{H}_1 \, dt_1 + \varepsilon^2 \int_0^T \left[\overline{H}_2 + \int_0^\tau \overline{H}_1 \, dt_2 \right] d\tau + \ldots \right]$$

$$\overline{H}_1 = e^{-H_o t} H_1(t) e^{H_o t} \tag{8}$$

And by the perturbation method for time-independent systems we can find the eigenvalues of B.

III. DYNAMIC EQUILIBRIUM

A dynamic equilibrium is a periodic solution of the system (1) with $\oint \underline{V}_o \, dt = 0$. The displacement vector $X(x,t)$ is defined by:

$$\frac{dX}{dt} = \underline{V}_o (X, t) \tag{9}$$

As a smallness parameter we introduce $\varepsilon^3 = \max |\underline{X}| / L$ (L = scale length of static equilibrium) and assume that frequency Ω of the dynamic equilibrium is very large $\Omega = 0 (1/\varepsilon)$. With these assumptions we obtain $\underline{V}_o = 0(\varepsilon^2)$, $\partial \underline{V}_o / \partial t = 0(\varepsilon)$. Writing the dynamic equilibrium in the form $\underline{B} = \underline{B}_o(x) + \underline{B}'_o(x, t)$, $p = p_o(x) + p'_o(x,t)$ $N = N_o(x) + \rho_o(x,t)$, from Eq. (1)

$$B'_o = 0(\varepsilon^3), \quad \beta'_o = 0(\varepsilon^3), \quad P'_o = (\varepsilon^2) \tag{10}$$

From this ordering $\partial \underline{V}_o / \partial t$ is the dominating term.

By linearization of Eq. (1) we obtain up to the order ε^2

$$N_o \left(\frac{\partial v_1}{\partial t} + \varepsilon^2 \left((\underline{v}_o \cdot \nabla) \underline{v}_1 + (\underline{v}_1 \cdot \nabla) \underline{v}_o \right) \right) + \rho_1 \varepsilon \frac{\partial \underline{v}_o}{\partial t} =$$

$$\underline{j}_1 \times \underline{B}_o + \underline{j}_o \times \underline{B}_1 - \nabla p_1$$

$$\frac{\partial \underline{B}_1}{\partial t} = \nabla \times (\underline{v}_1 \times \underline{B}_o) + \varepsilon^2 \nabla \times (\underline{v}_o \times \underline{B}_1)$$

$$\frac{\partial \rho_1}{\partial t} = - \operatorname{div} (N_o \underline{v}_1 + \varepsilon^2 \rho_1 \underline{v}_o)$$

$$\frac{\partial p_1}{\partial t} + \varepsilon^2 \underline{v}_o \cdot \nabla p_1 + \underline{v}_1 \cdot \nabla p_o + \gamma p_o \operatorname{div} \underline{v}_1 = 0 \qquad (11)$$

In the following we give a formal solution of the system, Eq. (11). It can be shown that in the dispersion relation the terms of order ε^2 are of no significance. They can thus be omitted in Eqs. (11)

$$N_o \frac{\partial^2 \underline{v}_1}{\partial t^2} + \frac{\partial}{\partial t} \rho_1 \varepsilon \frac{\partial \underline{v}_o}{\partial t} = F \underline{v}_1$$

$$\frac{\partial \rho_1}{\partial t} + \operatorname{div} N_o \underline{v}_1 = 0 \qquad (12)$$

F is the MHD stability operator of the static equilibrium. According to the basic theorem we require solutions of the form

$$\underline{v}_1 = \sum_{-\infty}^{-\infty} \underline{v}_n (x) \exp \left[i (n\Omega - \nu) t \right]$$

This leads to the system

$$\left[(\omega + n)^2 N_o + \frac{1}{\Omega^2} F \right] \underline{v}_n + (\omega + n) \varepsilon \frac{X_o}{2} \times$$

$$\operatorname{div} N_o \left(\frac{\underline{v}_{n+1}}{\omega+n+1} - \frac{\underline{v}_{n-1}}{\omega+n-1} \right) = 0 \qquad (13)$$

$$n = -\infty, \ldots 0, 1_1 \ldots + \infty, \qquad \omega = \frac{\nu}{\Omega}$$

For the velocity $\underline{V}_o(x_1 t)$ we have made the ansatz $\underline{V}_o = \Omega \underline{X}_o(x) \cos\Omega t$. To make all calculations self-consistent, a periodic solution of equations (1) should be found, but we consider the ansatz above as the first term in a Fourier expansion. By expansion with respect to ε we have

$$\underline{v}_n = \underline{v}_n^o + \varepsilon \, \underline{v}_n^1 + \varepsilon^2 \underline{v}_n^2 + \cdots$$

$$\frac{1}{\Omega^2} = \lambda_o(\omega) + \varepsilon \, \lambda_1(\omega) + \varepsilon^2 \lambda_2(\omega) + \cdots \qquad (14)$$

The second equation of (14) is the dispersion relation. As can be seen from Eq. (13), the system is invariant under the transformation $\omega \to -\omega$, $n \to -n$, $\underline{v}_n \to \underline{v}_{-n}$. $-\omega$ is also a solution of the dispersion relation. Therefore we have to look for real solutions of Eq. (14). The calculations of $\lambda_i(\omega)$ is straightforward, and we find

$$\lambda_o(\omega) = \frac{(\omega + N)^2}{\omega_k^2} \,, \quad \text{with } \omega_k^2 = -\frac{(\underline{v}_k, F\underline{v}_k)}{(\underline{v}_k, N_o \underline{v}_k)} \,, \quad N = \text{integer}, \quad (15)$$

and $\underline{v}_k$ are the eigenmodes of F;

$$\lambda_1(\omega) = \pm \frac{(\underline{v}_k, \frac{\underline{X}_o}{2} \, \text{div } N_o \underline{v}_k)}{(\underline{v}_k, F\underline{v}_k)} \,, \quad \text{if } \omega = \frac{M}{2} \text{ with } M = \text{integer};$$

$$= 0, \quad \text{if } \omega \neq \frac{M}{2} = 0; \qquad (16)$$

$$\lambda_2(\omega) = -\frac{\left(\underline{v}_k, \frac{\underline{X}_o}{2} \, \text{div } N_o \left[(G_{N-1}^{-1} - G_{N-1}^{-1})\, \underline{X}_o\right] \text{div } N_o \underline{v}_k\right)}{(\underline{v}_k, F \underline{v}_k)} \qquad (17)$$

with $G_{N\pm 1} = (\omega + N \pm 1)^2 \, N_o + \lambda_o F$.

The stability limit is given by the curve $1/\Omega^2 = \lambda_o(0) + \lambda_2(0)\,\varepsilon^2 + \cdots$ The result is, for $N = 0$,

$$-\frac{1}{2} \frac{(\underline{v}_k, \underline{X}_o \, \text{div } \underline{X}_o \, \text{div } N_o \underline{v}_k)}{(\underline{v}_k, N_o \underline{v}_k)} \, \varepsilon^2 \leq \frac{\omega_k^2}{\Omega^2} \leq \frac{1}{4} - \frac{(\underline{v}_k, \frac{\underline{X}_o}{2} \, \text{div } N_o \underline{v}_k)}{(\underline{v}_k, N_o \underline{v}_k)} \qquad (18)$$

corresponding to the stability conditions of Mathieu's equation.

IV. DISCUSSION

The zeroth-order system, $\left[(\omega + N)^2 N_o + \lambda_o F\right] \underset{N}{v}{}^o = 0$, is self-adjoint. Therefore the solutions $v_N^o = v_k$ can be real functions, and consequently all coefficients λ_o, λ_1, λ_2 are real for real ω.

As can be seen from the stability condition (18), the frequency Ω has to be larger than the growth $|\omega_k|$ of the instability, i.e., $-1/2 \, C_1^k (A/L)^2 \le \omega_k^2/\Omega^2$, where $C_1^k \approx 1$ and A = amplitude of $\underline{X}_o$. Parametric resonance occurs at $\Omega = 2\omega_k$. There is no stabilization if $\underline{v}_k$ and $\underline{X}_o$ are perpendicular to each other. The problem which has not been touched here is the calculation of $\underset{-o}{V}(x, t)$.

For technical reasons (large energy dissipation in the wall) it is not favorable to have time-dependent boundary values of the magnetic field. Therefore, it is necessary to look for nonlinear eigenmodes of the static equilibrium.

Acknowledgements:

The authors are indebted to D. Pfirsch for helpful discussions.

6.3 DYNAMIC STABILIZATION OF THE m = 1 INSTABILITY IN A BUMPY THETA PINCH

G. Berge
University of Bergen, Norway

ABSTRACT

We investigate the possibility of dynamic stabilization of slightly bumpy axisymmetric systems, using a general theory which can be regarded as a generalization of the well-known energy principle. The equilibrium is taken to be one in which there is a sharp boundary, with a surface current but no current in the main body of the plasma. An oscillatory external magnetic field then sets up oscillatory motion in the plasma column. The problem is made soluable by assuming the sharp boundary to have a prescribed motion. This motion will in general cause induced currents to flow in the plasma. A sufficient condition for dynamic stabilization of this system is obtained.

I. INTRODUCTION

Theoretical (Haas and Wesson, 1966;1967b)and experimental[†] (Bodin, McCartan, Newton, and Wolf, 1969) investigations of a linear (as opposed to a toroidal) high-β bumpy theta pinch show that it is always unstable. One possibility for its stabilization is dynamic stabilization (Weibel, 1960; Haas and Wesson,1967a;Troyon, 1967; Berge, 1969a, 1969b). Results from straight systems of this kind are relevant to large aspect ratio toroidal systems. In this connection the M and S systems (Meyer and Schmidt, 1958; Pfirsch and Wobig, 1966; Morse, Risenfeld, and Johnson, 1968; Lotz, Remy, and Wolf, 1964; Wolf, 1969b) are of special significance. An expansion of a toroidal system in the reciprocal aspect ratio results in a linear system in leading order. In addition, a linear system is much easier to handle mathematically than the toroidal one.

In the following analysis we first outline the model being studied. The m = 1 instability of this model is then investigated, and results are discussed.

II. MODEL.   A. Equilibrium.   We study a high-β axi-symmetric system where the plasma is confined to the region around the axis

---

[†] H.A.B. Bodin, A.A. Newton, G.H. Wolf, and J.A. Wesson, to be published in Physics of Fluids.

by a mainly axial magnetic field $(B_z \gg B_r)$. The axis is straight and the plasma assumed to be separated from the vacuum region by a sharp boundary carrying a sheet current. The equilibrium is the same as that investigated by Haas and Wesson (1967b). We also make use of their stability calculations, neglecting the stabilizing effect of finite infinitely conducting boundaries in the vacuum region. The stabilizing effect from a wall is small unless it is close to the plasma-vacuum interface. For obvious reasons this situation is not realized for highly compressed theta pinch plasmas. Cylindrical coordinates and the usual notation are used.

B. The oscillatory state. It is assumed that by controlling the external magnetic field, an oscillatory motion is set up around the equilibrium state just described. Although this has to be done experimentally by oscillating currents in external circuits, we assume that by doing this properly we are able to give the interface between plasma and vacuum a prescribed motion. This interface is assumed to be given by:

$$F(r,z,t) \equiv r - R(z,t) = 0 \tag{1}$$

where

$$R(z,t) = \overline{R}(z)\left\{1 + \varepsilon \sin \omega_s t \cos \alpha z\right\} = \overline{R}(z) + \tilde{R}(z,t). \tag{2}$$

We assume that $\varepsilon \ll 1$, and ω_s is the frequency of the oscillations, $\alpha = 2\pi/L$ where L is the wavelength in the axial direction. The magnetic field is assumed to remain tangential to the plasma-vacuum interface during the course of motion. The following solutions are obtained for the oscillatory motion:

$$\tilde{\underline{v}} = \varepsilon\left\{\omega_s \overline{R}S \cos \alpha z, \; 0, \; c_s^2 \alpha A Q \omega_s^{-1} \sin \alpha z\right\}\cos \omega_s t + 0(\varepsilon^2), \tag{3}$$

$$\tilde{\rho} = - \varepsilon \overline{\rho} A Q \cos \alpha z \sin \omega_s t + 0(\varepsilon^2), \tag{4a}$$

$$\tilde{p} = c_s^2 \tilde{\rho} + 0(\varepsilon^2), \tag{4b}$$

$$\tilde{\underline{B}} = - \varepsilon \, \overline{B}_{izo}\left\{\alpha \overline{R}S \sin \alpha z, \; 0, \; Q \cos \alpha z\right\}\sin \omega_s t + 0(\varepsilon^2). \tag{5}$$

The symbols are defined as follows: $Q = x_0 \, J_0(x)/J_1(x_0)$, $S = J_1(x)/J_1(x_0)$, J_0 and J_1 are the zero and first order Bessel functions, $A = (1-a^2)^{-1}$, $a = \alpha \, c_s \omega_s^{-1}$, $x_0 = k\overline{R}$, $x = kr$,

$$k = \frac{\omega_s}{C_s}\left[\frac{1-ba^2}{A+b}\right]^{\frac{1}{2}}, \quad b = \frac{2}{\gamma}\frac{1-\beta_o}{\beta_o}, \quad \beta_o = \frac{2\bar{p}}{\bar{B}^2_{vzo}}, \quad C_s^2 = \frac{\gamma\bar{p}}{\bar{\rho}},$$

and the barred quantities refer to the equilibrium. Subscripts i and v refer to the internal plasma region and the vacuum region respectively.

III. STABILITY. The stability analysis of this model is based on results derived elsewhere (Berge, 1969a, 1968).   The problem is to investigate the sign of [†]

$$\delta\overline{W} = \delta\overline{W}_O + \delta\overline{W}_D + \delta\overline{W}_E \tag{6}$$

where

$$\delta\overline{W}_O = -\omega_s^{-2}\int \underline{\xi}_o^*\cdot(\varepsilon^{-2}\overline{F}_e + \tilde{F}_2)(\underline{\xi}_o)d\underline{r}, \tag{7a}$$

and

$$\delta W_D = \int \bar{\rho}^{-1}|F_1'(\underline{\xi}_o)|^2 d\underline{r}. \tag{7b}$$

According to the previous results (Berge, 1968) (Appendix H, Sec. b), $\delta W_E \geq 0$, and we can obtain sufficient conditions for stability with this term omitted. Analysis shows that for the case m = 1 (and $\partial/\partial r = 0$ for the perturbations) δW_D minimizes to zero for $\xi_z = 0$. Moreover the most unstable modes are those for which $\nabla\cdot\underline{\xi}_o = 0$ and $\partial\xi_r/\partial z$ is small or of the order ε ($\underline{\xi}_o = \{\xi_r, \xi_\theta, \xi_z\}$). After a considerable amount of algebra involving expansions in small parameters, we arrive at the following expression

$$\delta\overline{W} \geq 2\pi\int\left\{a\left|\frac{d\xi_r}{dz} + \frac{b}{a}\xi_r\right|^2 + (c - \frac{b^2}{a})|\xi_r|^2\right\}\bar{B}^2_{vzo}\,dz, \tag{8}$$

where

$$a = (2-\beta_o)\overline{R}^2, \quad b = 2(1-\beta_o)\overline{R}\frac{d\overline{R}}{dz},$$

$$c = 2(1-2\beta_o)(1-\beta_o)(\frac{d\overline{R}}{dz})^2 + (\frac{\partial\tilde{R}}{\partial z} + \varepsilon\,\overline{R}\,\bar{B}^{-1}_{vzo}\frac{\partial}{\partial z}\tilde{B}_{vz1})^2 \tag{9}$$

$$+ \varepsilon\;2\overline{R}\,\bar{B}^{-1}_{vzo}\,\tilde{B}_{vz1}\frac{\partial^2\tilde{R}}{\partial z^2}.$$

---

† The notation used here is the same as that in Berge(1968).

From Eq. (8) we obtain the following sufficient condition for MHD
stability

$$\langle \tilde{c} \rangle - \langle \frac{\tilde{b}^2}{c} \rangle > 0 .$$

(10)

where $\langle \rangle$ denotes averaging over z, $-$ averaging over t.   From the
pressure balance condition we obtain in leading order:

$$\tilde{B}_{vz1} = - \bar{B}_{vzo} Y \cos \alpha z \sin \omega_s t,$$

(11)

where

$$Y = Q \left[1 - \beta_o + \frac{\gamma \beta_o}{2} A \right] .$$

Let $I = \left[- \frac{\pi}{\alpha}, \frac{\pi}{\alpha} \right]$ be an interval on the z-axis.   We define:

$$\bar{R} = R_o \left[1 + \delta(1 + \cos \alpha z) \right] , \quad z \in I \quad ;$$

$$\bar{R} = R_o, \quad z \notin I.$$

(12)

This gives us a single bulge with amplitude $2\delta R_o$.   By averaging
condition (10) over the bulged region we obtain the stability criterion:

$$\frac{\varepsilon^2}{4} \left[Y^2 + 1 \right] - \left[\frac{\beta_o (1 - \beta_o) (3 - 2\beta_o)}{2 - \beta_o} \right] \delta^2 > 0 ,$$

(13)

IV. DISCUSSION.   For parameter values relevant to the experimental
results[†] ($Y \approx 2$, $\beta_o \approx 0.5$) condition (13) becomes $\varepsilon > 0.5 \delta$.
This can explain the observed effect.

By considering a periodic structure rather than a single bulge
one obtains the same condition (13).   One would think that this case
had some relevance to large-aspect ratio M and S systems. Realistic
values of Y are of the order 2-3, and one can easily estimate the
amount of dynamic stabilization required to stabilize a given δ
(Morse, Risenfeld, and Johnson, 1968).

The present scheme may also be used to produce "dynamic
equilibrium" where no equilibrium would otherwise exist (Wolf and
Berge, 1969).

[†]H.A.B. Bodin, E.P. Butt, J. McCartan, and G.H. Wolf
to be published.

In the context of a fusion reactor the application of dynamic stabilization will become an economic problem. This has not yet been properly considered. Dynamic stabilization may, however, also prove useful in cases where one has to pass through an unstable situation in order to build up a configuration which finally is stable. A more detailed version of this work has been published recently (Berge, 1970).

DYNAMIC STABILIZATION EXPERIMENTS IN Q-DEVICE,
DISCHARGES, AND BEAM-PLASMA SYSTEMS

7.1 PRELIMINARY RESULTS ON THE DYNAMIC STABILIZATION OF THE "DRIFT-DISSIPATIVE" INSTABILITY

M. W. Alcock and B. E. Keen
UKAEA, Research Group, Culham Laboratory,
Abingdon, Berkshire, England

ABSTRACT

Preliminary results are presented which show that the "drift-dissipative" instability in helium and hydrogen afterglow plasmas is dynamically stabilized by an oscillating magnetic field B_θ of frequency ω. Oscillations at the fundamental frequency $(\omega_o \sim 4 \text{ kHz})$ are suppressed with $B_\theta/B_z \sim 1\%$; some signal at $2\omega_o$ remains. Larger-amplitude drives were required to suppress this second harmonic. Experiments have also been performed to determine how the critical field for stabilization B_θ^{crit}, varies as a function of the oscillating frequency ω (in the range 8 kHz to 100 kHz), both for suppression of the fundamental instability frequency ω_o, and for its second harmonic at $2\omega_o$.

I. INTRODUCTION

Recently, there has been considerable interest in the stabilization of various types of drift instabilities present in a magneto-plasma, due to the anomalous diffusion of plasma across the containing magnetic field associated with these instabilities. Among the possible methods of stabilization is that suggested by Ivanov and Teichmann (1969). In this case, a high frequency (h.f.) oscillating magnetic field B_θ at a frequency $\omega \gg \omega_o$ (the instability frequency) is applied to the plasma in the azimuthal direction perpendicular to the containing axial magnetic field, B_z. When B_θ reaches an amplitude such that $B_\theta/B_z > \omega/k_\theta v_T$ the instability should be suppressed. Here, k_θ is the azimuthal wave number and $v_T = (T_e/m)^{1/2}$ is the electron thermal velocity.

The physical mechanism of stabilization is as follows:

The density perturbations in the plasma form flutes ($k_z \ll k_\theta$) oriented along the direction of the d.c. magnetic field B_z, and the distance between these flutes is $2\pi/k_\theta$. During the oscillation period $2\pi/\omega$ of the instantaneous field line of the total magnetic field ($B_z + B_\theta \cos \omega t$), the density perturbations can be regarded as time independent since $\omega \gg \omega_0$. If this instantaneous field line intersects two neighboring density maxima, electrons can traverse these density maxima, and thus "short-out" the instability. Consequently, the mechanism of stabilization is based upon the change of spatial distribution of electrons during this time period, $2\pi/\omega$.

In particular, for the drift dissipative instability, the effect of this h.f. field was taken into account (Ivanov and Teichmann, 1969) in the linearized kinetic equation approach. The problem was considered in the low-frequency approximation ($\omega_0 \ll \Omega_i = eB_z/Mc$), and both an electron-neutral ν_e and an ion-neutral ν_i collision frequency were taken into account. The following complex dispersion relation quadratic in the collision frequency ω_0: was obtained

$$\omega_0^2[1 + \omega^* \omega_s / k_z^2 D_e \omega \beta^{2/3}] + i\omega_0[k_z^2 D_e + \nu_i + \omega_s] - i\omega^* \omega_s - k_z^2 D_e \nu_i = 0 \tag{1}$$

where

$$\omega^* = ck_\theta T_e \left(\frac{dn}{dr}\right)/enB_z \quad , \quad \omega_s = k_z^2 \Omega_e \Omega_i / k_y^2 \nu_e \quad ,$$

$\beta = [k_\theta^2 D_e B_\theta^2 / 4 \, B_z^2 \omega] \gg 1$, $D_e = T_e/m\nu_e$, and Ω_e and Ω_i are the electron and ion cyclotron frequencies, respectively. In this case, the oscillation frequency ω_0 was changed very little by the applied frequency ω, but the growth rate $\gamma (= \text{Im } \omega_0)$ was reduced to:

$$\gamma = \frac{(\omega^* \omega_s)^3}{(\omega_s + \nu_i)^3 \omega k_z^2 D_e \beta^{2/3}} - \frac{k_z^2 D_e \nu_i}{(\omega_s + \nu_i)} \tag{2}$$

and stabilization occurs ($\gamma \leq 0$) when:

$$\left(\frac{B_\theta^c}{B_z}\right) \approx \left[\frac{16(\omega^*)^9 \omega_s^3}{\omega(k_z^2 D_e)^8 \nu_i^3}\right]^{1/4} \tag{3}$$

Equation (3) shows that the critical value for stabilization, B_θ^c, is proportional to $(1/\omega)^{1/4}$.

III. EXPERIMENTS

The experiments were performed in an afterglow plasma, which was produced by using a short pulse of electrons emitted from a hot cathode to ionize a neutral gas (helium or hydrogen). Typical initial densities $n_0 \sim 10^{10}$ cm^{-3} were obtained, and an average electron temperature $T_e \sim 1000^\circ$ K was measured with a single Langmuir probe. The instability was observed as an oscillating sinusoidal potential imposed on the exponentially decaying background signal. The dispersion diagram (ω - k_z) was obtained under various conditions and this showed the instability to be the drift-dissipative instability (Alcock and Keen, 1970). The experiments on dynamic stabilization were carried out with the instability at ~ 2.5 kHz in a magnetic field $B_z = 225$ G and with the longitudinal wavelength $\lambda_z \sim 150$ cm. The oscillating magnetic field, B_θ, was applied to the plasma, near a maximum in the longitudinal amplitude, from four coils spaced at equal intervals around the periphery of the tube oriented such that each coil produced an in-phase oscillating azimuthal field.

IV. RESULTS

Figure 1 shows the effect of this applied oscillation taken at a frequency of 20 kHz. Figure 1(a) shows the signal obtained with $B_\theta = 0$; Fig. 1(b) shows the effect when $B_\theta/B_z \sim 0.005$; and Fig. 1(c) shows the effect when the amplitude is increased to $B_\theta/B_z \sim 0.01$. It is seen that suppression of the instability at $\omega_0 = 2.5$ kHz is achieved for $B_\theta/B_z \sim 1\%$, but that at this point a signal at $2\omega_0 \approx 5$ kHz has appeared. If the amplitude B_θ is further increased this instability at $2\omega_0$ can be suppressed with $B_\theta/B_z \sim 6.5\%$ at this applied frequency ($\omega_c = 20$ kHz). The critical field required for suppression, B_θ^c, as a function of applied frequency, ω, is shown in Fig. 2, both for suppression of the instability at frequency ω_0 and that at $2\omega_0$. Also, shown on this diagram as the dashed curve $[B_\theta^c \propto (1/\omega)^{\frac{1}{4}}]$ is that predicted by the theory. It is seen that the magnitude of this field, B_θ^c, falls with increasing frequency, ω, as predicted by theory, but the variation appears to be more rapid than the $B_c \propto (1/\omega)^{\frac{1}{4}}$ shown in the diagram.

It was checked that the suppression was not due to possible local heating or plasma production effects caused by the h.f. source. This check was effected by using an oscillating magnetic field, B_z, applied in the axial direction. No suppression effects were observed even with fields an order of magnitude higher in amplitude. Therefore, it is inferred that the mechanism of suppression, in this case, is the method suggested by Ivanov and Teichmann. Experiments

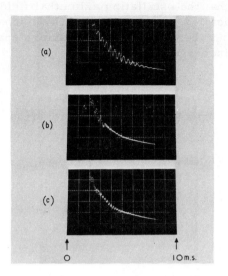

Fig. 1. Floating probe output showing the instability at 2.5 kHz subjected to an oscillating magnetic field, B_θ, such that (a) $B_\theta/B_z = 0$; (b) $B_\theta/B_z \sim \frac{1}{2}\%$; and (c) $B_\theta/B_z \sim 1\%$.

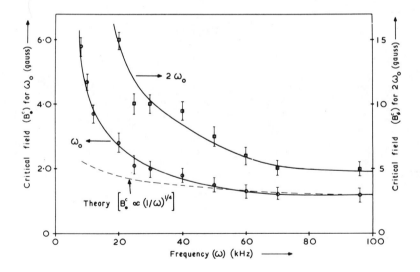

Fig. 2. The measured critical magnetic field, B_θ^c, plotted against the applied frequency for suppression of the instability at ω_o (left hand scale) and at $2\omega_o$ (right hand scale). The dashed curve shows the theoretically predicted variation $[B_\theta^c \propto (1/\omega)^{\frac{1}{4}}]$.

are continuing, so that the oscillating azimuthal field can be applied from an axial conductor down the center of the discharge tube, or alternatively, by a number of parallel conductors spaced around the outside of the discharge tube.

7.2 SUPPRESSION OF DRIFT INSTABILITY
BY RF ELECTRIC FIELD*

Y. Nishida[†], M. Tanibayashi and K. Ishii

Institute of Plasma Physics, Nagoya University, Nagoya, Japan

ABSTRACT

Suppression of the drift instability is achieved by applying a homogeneous rf electric field in a dc discharge diffused plasma. The instability amplitude is reduced by more than 25 db. The suppressing rf frequency depends on neutral pressure but is nearly independent of plasma density. These phenomena are discussed based on a model of high-frequency plasma conductivity.

I. INTRODUCTION

Suppression of the drift-wave instability by externally applied rf electric field has been reported (Nishida, Tanibayashi, and Ishii, 1970). It was shown experimentally that in certain frequency ranges of the applied rf field, a strong coupling zone exists where drift waves were considerably suppressed. In this paper we discuss the experimental results on parametric dependences for the coupling zone.

II. EXPERIMENTS

The experiments are performed with the TPD-I machine of Nagoya University (Takayama, Otsuka, Tanaka, Ishii, and Kubota, 1967). The plasma has a density ranging from 2×10^9 to $6 \times 10^{11} cm^{-3}$ with a helium pressure of 1 to 3×10^{-3} Torr. The plasma column is 90 cm long and 2.0 cm in diameter, with uniform magnetic field B_0, from 0.7 to 3.0 kG. Rf electric fields of frequency 5MHz to 50 MHz and amplitudes up to 26 V peak-to-peak, are applied along the plasma column. A radially movable Langmuir probe is used to measure the plasma parameters; to detect the drift wave we bias the probe to collect ion saturation-current. The wave signal so obtained is analyzed on the spectrum analyzer, as well as on a

---

*Work performed under the Collaborating Research Program at Institute of Plasma Physics, Nagoya University, Nagoya, Japan.

[†] Permanent address: Tohoku University, Sendai, Japan.

dual-beam cathode-ray oscilloscope.

Spontaneously excited low-frequency (25 to 80 kHz) oscillations
are identified as the drift instabilities, after making corrections for
the finite ion inertia and for the plasma rotation due to radial
electric field[*].

For rf frequency f_{rf} lower than the ion plasma frequency f_{pi}
(~ 2 MHz), based on plasma density at the axis, no change in drift
wave amplitude is observed.   For f_{rf} several times higher than
f_{pi} ($f_{pi} < 10$ MHz $\lesssim f_{rf} \lesssim 40$ MHz) and at an appropriate amplitude
(about 10-27 V peak-to-peak), the amplitude of the drift wave is
found to be attenuated by more than 25 dB, and its frequency shifts
slightly.

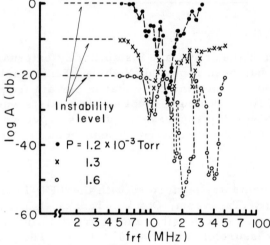

Fig. 1. Instability amplitude versus rf field frequency for three
different neutral-pressure conditions. $B_0 = 886$ G. Closed circles:
Id = 4.0 A; crosses: Id = 3.0 A; and open circles: Id = 100 mA.

The strong coupling zone for drift-wave suppression has a neutral pressure dependence as shown in Fig. 1. The coupling zone
shifted to the higher-frequency range and broadens slightly, when
neutral pressure is increased. The frequencies in the coupling zone
are several times higher than the electron-neutral collision frequencies (about $4 \sim 5$ MHz) obtained from published data (Brown,
1959) and with the use of measured electron temperature. Figures
2 and 3 show that the coupling zone hardly depends on the plasma

---

[*]Y. Nishida and Y. Hatta, to be published.

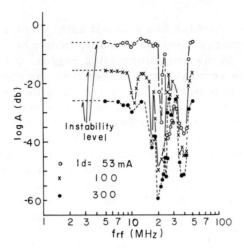

Fig. 2. Instability amplitude versus rf field frequency for three different discharge-current conditions. B_0 = 886 G, and p = 1.6 × 10^{-3} Torr.

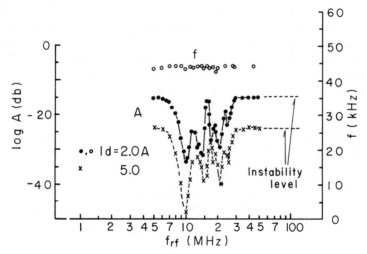

Fig. 3. Instability amplitude and frequency versus rf field frequency. B_0 = 886 G, and p = 1.3 × 10^{-3} Torr.

density, which determines the ion- or electron-sheath resonance frequency, (The ion-sheath resonance frequency is nearly equal to the frequencies in the coupling zone.) We cannot discuss the characteristics of the fine structure of the coupling zone, as we do not have sufficient data yet.

III.   DISCUSSION

The plasma in our machine has a small axial electron beam in addition to the lower energy electron stream, even when the target electrode is floating. In the experimentation region, the electron beam, whose energy is several times higher than the mean electron temperature, raises the ionization gradually along the axial flow direction. The electron beam is thought to be produced and accelerated by the electric potential difference between anode and the last floating electrode. This beam may not be thermalized, because the electron-neutral-collision mean free path is nearly the same as, or longer than, the distance between grid and target electrode. Electron collisions with other particles are negligibly small when compared with electron-neutral collisions. Ion-neutral- and ion-ion-collision mean free paths, on the other hand, are over ten centimeters. Therefore, we may further conclude that the drift wave is excited by an axial electron stream in addition to collisional effects. Ion collisions have stabilizing effects on this wave.

Applied rf fields will modulate the electron beam and, therefore, the ionization rate. If the modulation frequency ω_{rf} is nearly equal to that of the standing wave (whose phase velocity is the same as the electron beam velocity) between exciting grid and target electrode the rf plasma conductivity may be enhanced and the rf power is fed into plasma resonantly. The complex rf conductivity σ is obtained with the use of the equations of motion and continuity for electrons:

$$\sigma = \frac{N_o e^2}{m_e} \left[\nu - V_o \lambda + \frac{kV_o \nu}{\omega_{rf} - kV_o} \right.$$

$$\left. + j\left(\omega_{rf} + \frac{\lambda V_o \nu}{\omega_{rf} - kV_o} \right) \right] \frac{1}{(\omega_{rf} - kV_o)^2 + \nu^2} ,$$

(1)

where V_o is the electron beam velocity along the plasma column, k the wave number, ν the electron-neutral-collision frequency, $\lambda = -\partial \ln N_o / \partial x$, N_o the steady-state beam density, and the x-direction is along the plasma axis. Other notations are the standard ones. If the beam velocity V_o is zero, the expression for σ reduces to that obtained by Margenau (1964) under the assumption that the collision frequency is almost independent of electric-field intensity.

Equation (1) shows that if $\omega_{rf} \simeq kV_0$, the conductivity increases resonantly and hence the rf power is strongly absorbed. If we choose the beam energy as ~20 eV, (about 4 ~ 5 times larger than the thermal energy of electrons) and wave number k as π/L (L is the distance between grid and target electrode), then $kV_0 \simeq 10^7$ rad/sec. Although this frequency is several times lower than the frequencies of the coupling zone, the discrepancy does not seem serious because the wavelength may be shorter than the distance between the electrodes. This model also shows that the characteristic coupling-zone structure is independent of plasma density. The mechanism for suppressing the drift-wave instability is not clear and is being investigated.

ACKNOWLEDGMENT

It is a pleasure to thank Prof. K. Takayama, Prof. Y. Hatta and Dr. T. Sato for their discussions and encouragements.

7.3 DYNAMIC STABILIZATION OF A
TWO STREAM ION INSTABILITY

J. F. Decker and A. M. Levine
Bell Telephone Laboratories, Inc., Whippany, New Jersey   07981

ABSTRACT

In a certain pressure range, the Whippany ECR collisionless plasma has been observed to be unstable to a two stream ion instability The frequency spectra of the instability has been related to the double humped ion distribution function measured with an energy analyzer. By modulating the power source of the plasma, a time-dependent variation of plasma parameters was introduced. This has been used to stabilize the microinstability. As the frequency of the modulation was varied relative to that of the instability, regions of stabilization and destabilization were observed.

I. INTRODUCTION

Among the techniques that have been proposed for dynamic control of plasma instabilities are feedback stabilization, high frequency stabilization, and asynchronous quenching. In this note we present preliminary results on the stabilization of a microinstability by low-frequency modulation of the distribution function.

II. DESCRIPTION OF APPARATUS

A schematic of the Whippany ECR plasma device and the axial magnetic field profile are shown in Fig. 1. Up to 3 watts of microwave power (4 GHz) were absorbed in a cyclotron resonance region located at the maximum of a 3.1:1 mirror field. The plasma column was terminated ~1 meter from the microwave source [†] by a floating plate. An electrostatic energy analyzer measured the energy of ions passing through a 4 mm hole in the end plate. The device was also instrumented with radially and axially movable Langmuir probes. Typical operating conditions for these experiments are : $10^8 < N < 10^9$ cm^{-3} and $T_e \sim$ 5-6 eV.

[†] J.F. Decker and C. D'Amico, submitted to Review of Scientific Instruments.

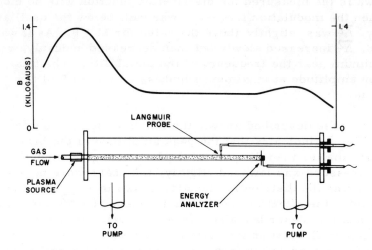

Fig. 1. Schematic drawing of Whippany ECR plasma device
and magnetic field profile.

III.   TWO STREAM ION INSTABILITY

In the pressure range 5×10^{-6} - 5×10^{-5} torr of argon, a large
amplitude, long wavelength, 2-3 KHz oscillation was observed.
Simultaneously, the ion energy distribution was found to consist of
two components: a low energy part and a beam. Beam ions originate
in the source and charge-exchange with background neutrals to produce the low-energy component. For a pressure of 2×10^{-5} torr the
energy of the beam varied from 14-18 eV as the microwave source
power was increased from 1 to 3 watts. For these experiments, the
background pressure was adjusted so that the two components were
nearly equal in amplitude at the end of the column. The frequency of
the instability is determined by the machine length and corresponds
to that predicted for the two stream ion instability assuming that the
machine length is a half wavelength. A detailed discussion of this
instability will be presented in a future communication.

IV.   EXPERIMENT

When the source power P was modulated with a square wave a
strong effect was observed on the instability. We have measured
the mean square of the potential fluctuations A^2 using a wide band
(5 Hz - 4 MHz) rms. voltmeter. In Fig. 2, we show A^2 vs. modulation frequency Ω at a fixed modulation amplitude M = $\Delta P/P$ of 12%.

Also shown is the measured ion distribution function with no modulation. When the modulation frequency was well below the oscillation frequency, $\overline{A^2}$ was slightly above the value for M = 0. As Ω was increased, $\overline{A^2}$ increased slowly but then decreased suddenly, reaching a minimum near the frequency of the instability. The total fluctuation amplitude at maximum suppression was 40% of the M = 0 value.

For the case described above, the ion distribution function shown in Fig. 2b indicates that the peak amplitude of the beam component was smaller than the peak of the low energy component. When the pressure was lowered slightly, the peaks were nearly equal producing the distribution function shown in Fig. 3b. Under this condition, the effect of modulation frequency on A^2 was quite different than that described above (see Fig. 3a). As before, there was a minimum fluctuation intensity at about 40% of the M = 0 value occurring for a modulation frequency close to the instability frequency. But, in this case, when Ω was increased slightly above that point, there was a sudden increase in A^2, reaching a maximum 60% above the unmodulated value.

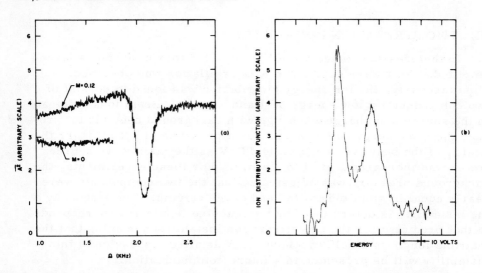

Fig. 2. (a) Mean square potential fluctuations vs. modulation frequency for M = 0 and M = 0.12. (b) Corresponding ion energy distribution function.

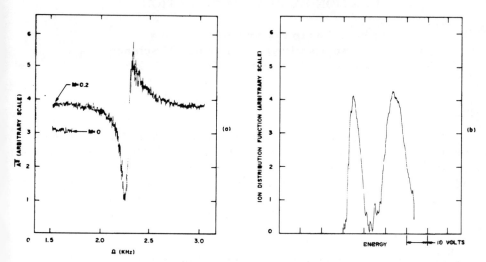

Fig. 3. (a) Mean square potential fluctuations vs. modulation frequency for M = 0 and M = 0.2. (b) Corresponding ion energy distribution function.

V. DISCUSSION

A theory describing the effect of time-dependent distribution functions on microinstabilities has been given by Dobrowolny, Engelmann and Levine (1969). Applying this theory to the two-stream ion instability, one finds significant stabilization effects for modulations in beam velocity of the order of a few per cent. Detailed theoretical curves are currently being numerically calculated. It is interesting to note that density modulations are much less effective in stabilizing this mode. Experimentally, the beam energy depends on the microwave power. Thus, the applied modulation is varying (among other things) the beam velocity. It is too early to state whether this can account for all of the effects observed.

In conclusion, the results presented here raise the possibility that microinstabilities may be stabilized by the appropriate modulation of the distribution function.

Acknowledgments

The authors gratefully acknowledge the technical assistance of C. D'Amico. We also thank C. Bateman and L. Dozier for aid in numerical computations.

7.4 SUPPRESSION OF LONGITUDINAL ELECTRON OSCILLATION BY BEAM MODULATION

Yoshiharu Nakamura and Shoji Kojima
Department of Applied Physics, Faculty of Science,
Tokyo University of Education, Otsuka, Tokyo

I. INTRODUCTION

Control of beam-plasma instability by beam modulation has been reported by several authors. Kornilov et al.(1966) found that a low frequency oscillation, brought about by a high frequency one, was suppressed by modulating a beam at this frequency. Berezin et al. (1969) found that the beam-generated oscillation having a broad spectrum peaked at 880 MHz was controlled by beam modulation at two different frequencies.

The present experiment studies suppression of beam-excited longitudinal electron oscillations by beam density modulation in an axially and transversely finite beam-plasma system. The modulation frequency is in the vicinity of the frequency of the oscillation, neighbouring modes, or higher harmonics.

II. EXPERIMENT

In a glass tube 30 cm in diameter and 50 cm in length, an electron gun and a beam repeller were set up. A magnetic field of 44 G was applied along the axis of the tube. The gun consisted of an indirectly heated oxide cathode of 10 mm diameter, a control grid and a mesh anode. The distance between the anode and the repeller, D, was 17 cm. The anode voltage was 90 V and the beam current I_b was controlled by the grid. Modulation was produced by applying a sinusoidal voltage of frequency f_m to the grid. Argon gas was used with a pressure of $10^{-4} - 10^{-3}$ Torr. Oscillation signals were detected by a movable coaxial probe which was at floating potential and displayed on a spectrum analyzer. The plasma was generated by the beam; density and electron temperature as measured by Langmuir probe were $10^8 - 10^9/cm^3$ and 2 eV, respectively.

Oscillations were observed only when the repeller voltage V_r was about one half of, or larger than, the anode voltage V_a. From the measurement of current flowing to the repeller it was found that secondary electrons were emitted and went back to the gun when V_r was half of V_a and the beam was reflected when V_r was larger than V_a. The reverse electron flow probably caused feedback (Fedorchenko, Muratov, and Rutkevich, 1964), i.e. a signal amplified in the beam-plasma system is again applied to the beam input. As a result the system becomes a self-sustained high-frequency oscillator. The voltage V_r was set equal to V_a so that the beam was reflected. Standing electrostatic waves were excited between the anode and the repeller. The frequency f_n of the n-th mode of oscillations is expressed (Kawabe, 1966) by $(V_p/2D)n$, where V_p is the phase velocity.

III. RESULTS

When $V_a = 90$ V , $I_b = 320$ μA and $p = 9 \times 10^{-5}$ Torr, there existed only one oscillation having a frequency f_3 (37.4 MHz) corresponding to the n = 3 mode. On this oscillation, beam modulation effects were studied as follows:

1) When the modulation frequency f_m approached 37.4 MHz, the amplitude of this oscillation, f_3 , decreased and at the same time that of the signal at f_m increased. This phenomenon is quantitatively explained by consideration of the solution of a forced Van der Pol equation.[†]

2) When f_m changed to 50 MHz which corresponded to the frequency of the n = 4 mode, similar suppression of oscillation f_3 and enhancement of the signal at f_m were observed. This n = 4 oscillation, however, was not observed without changing discharge parameter. The same phenomenon was observed when f_m was nearly equal to 26 MHz corresponding to the n = 2 mode.

3) With f_m changed to 75 MHz or 112 MHz, i.e. higher harmonics of f_3, the subharmonic oscillation $(1/2)f_m$ or $(1/3)f_m$ appeared at the expense of the energy of oscillation f_3, whose frequency was nearly equal to $(1/2)f_m$ or $(1/3)f_m$.

In these cases, the spontaneous oscillation having spectral width of about 500 kHz was suppressed, but other oscillations were enhanced. The enhanced oscillations had narrow spectral width because they were driven by an external signal and the correlation time should be much longer than that of spontaneous oscillations.[*] The non-linear low frequency oscillation may be also suppressed by this narrowing of spectral width (Kornilov et al.1966).

When two or three oscillations of different modes exist together, all of them can be suppressed by applying modulation whose frequency coincides with one of the frequencies.

[†] Y. Nakamura, to be published in Journal of Physical Society of Japan.

[*] S. Kojima, A. Itakura, and Y. Nakamura, to be published in Journal of Physical Society of Japan.

7.5 NONLINEAR PERTURBATIONS AND ELECTRON SCATTERING IN A BEAM PLASMA

J. E. Walsh, D. Bancroft, [†] R. W. Layman, and P. B. Lewis
Department of Physics and Astronomy
Dartmouth College, Hanover, New Hampshire 03755

ABSTRACT

The scattering of electrons by fluctuations, out of an electron beam in a beam plasma, is measured experimentally. Evidence for a nonlinear effective inverse mean free path which depends upon the intensity and width of the fluctuations is presented. We find that some control over the scattering process can be obtained when external oscillators of the natural oscillations are used to perturb the natural fluctuation spectrum. It is possible to both enhance and decrease velocity space diffusion in this way.

I. INTRODUCTION

A modified theory of Brownian motion, in which allowance has been made for a frequency-dependent friction, has been developed by Kubo (1966) and Mori (1965). Here we will explore an application of this theory to the dynamic stabilization of high-frequency electron plasma waves in a magnetized beam plasma. The fluctuating force in the model is taken to be the electric field of the instability. Although the usual assumptions of Brownian motion theory are not rigorously met in such a system it is a reasonable approximation when the correlation time of the fluctuations is short in comparison to the overall beam relaxation time.

The experimental results were obtained in a beam-formed hydrogen plasma whose density was $10^9 - 10^{10}$ cm^{-3}. Electron temperatures ranged between 10 and 20 eV, and the beam energy was 1 keV. The spectrum of fluctuations was observed with electrostatic loops and a wide-band spectrum analyzer. Both plasma and beam electron flux distributions were measured in a separately pumped retarding grid energy analyzer. A schematic of the machine is shown in Fig. 1.

[†] Permanent address: Colby College, Waterville, Maine.

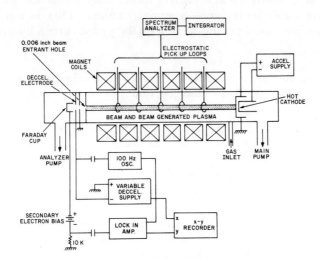

Fig. 1. Schematic layout of the apparatus.

II. APPLICATION OF RANDOM-WALK THEORY

The application of the theory to the random walk of electrons is straightforward. One begins with a particle equation of motion,

$$\dot{v} + \int_0^t \Lambda(t - t')\, v(t')dt' = -\frac{e}{m}\, E \tag{1}$$

where allowance has been made for a frequency-dependent friction,

$$\lambda(\omega) = \frac{e^2}{m^2 \overline{v}^{\,2}}\; \overline{E^2}_\omega \,. \tag{2a}$$

$$\lambda(\omega) = \int_0^\infty \Lambda(\tau) e^{i\omega\tau} d\tau \tag{2b}$$

Equation (2) follows from Eq. (1) when the fluctuation process represented by $E^2\omega$ is stationary. We might expect, however, that Eq. (2) would be a reasonable approximate expression for the collision frequency in the presence of unstable modes provided the time variation of the mean-square plasma electron velocity v^2 is slow in comparison to the characteristic fluctuation frequencies. We have found this to be the case. Furthermore, in applying this model to

the beam-plasma system we are tacitly assuming that a one dimen-
sional model is a sufficiently good approximation. This is not unreasonable since we are working with axially propagating longitudinal
modes.

When a finite number of (initially unstable) modes with moderate
width exist in the plasma, $\lambda(\omega)$ may be expressed as

$$\lambda(\omega) = \frac{\alpha}{\pi} \sum \frac{I_j \Gamma_j}{(\omega - \omega_{jo})^2 + \frac{\Gamma_j^2}{4}} \tag{3}$$

where I_j is the intensity, ω_{jo} the center frequency and Γ_j the width
of the jth mode. The coefficient α which contains v^2 is known from
a measurement of the distribution function.

The reaction of the fluctuations back on the beam are described
by the average collision frequency, $\overline{\lambda}$, seen by the beam electrons,

$$\overline{\lambda} = \frac{\alpha}{\pi} \sum_j I_j \frac{\Gamma_j}{\omega_{jo}^2} \tag{4}$$

A comparison of the measured intensity necessary to achieve an e-
folding of the beam flux with the value calculated from Eq. (4) is given
in Fig. 2. We find agreement over three orders of magnitude.

The collision frequency Eq. (3) which is nonlinear, will now be
introduced into the equation for the ith mode:

$$\dot{E}_i - (i\omega_{io} + \gamma_i)E_i - \int_0^t \Lambda(t - t')E_i(t')dt' = 0 \tag{5}$$

The growth rate γ_i is determined from the linearized dispersion
relation. An approximate equation for the intensity of the ith mode
is obtained by recognizing that the ith oscillator has a damping $\lambda(\omega)$
evaluated over a range of frequencies near ω_{io} . The equation for
the intensity I_i is then

$$\dot{I}_i - \gamma_i I_i + \beta_{ii} I_i^2 - \sum_j \beta_{ij} I_i I_j = 0 , \tag{6}$$

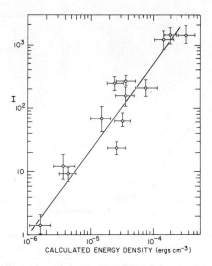

Fig. 2.   Typical data showing the dependence of I_1 the natural mode I_2 the driven mode.

an equation of the usual form (Landau and Lifshitz, 1959; Lashinsky, 1969; Wong and Hai, 1969). The coefficients follow from Eq. (3),

$$\beta_{ii} = \frac{\alpha}{4\pi\Gamma_i} \tag{7}$$

and

$$\beta_{ij} = \frac{\alpha}{\pi} \frac{\Gamma_j}{\Delta_{ij}^2 + \frac{\Gamma_j^2}{4}} \tag{8}$$

where

$$\Delta_{ij} = \omega_{io} - \omega_{jo} \tag{9}$$

In writing Eq. (7) we have assumed that the oscillator sees its own line width, i.e. there is a self energy term of order 2Γ. The coefficients β_{ij} are similar to those which occur in the Briet-Wigner (Davydov, 1965) cross-section which describes scattering in the presence of resonances. We will now apply these results to the dynamic stabilization of an unstable mode by a driven (stable) mode at higher frequency. The natural mode I_1 is easily expressed in terms of the driven mode I_2,

$$I_1 = \frac{4\pi\Gamma_1\gamma}{\alpha} - \frac{4\pi\Gamma_1\Gamma_2 I_2}{\Delta_{12}^2 + \Gamma_2^2} \tag{10}$$

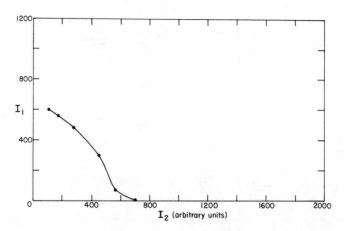

Fig. 3.   Typical data showing I_1 vs. I_2 for strong turbulence.

Typical plots of I_1 vs I_2 are shown in Fig. 3.   When the intensities
are in the general range of those in the figure the coefficient preceding
I_2 ranges between .05 and .5 in agreement with Eq. (10).   Some care
must be used in the construction of the coefficients in Eq. (10) since in
some cases the natural modes are actually a group of closely spaced
narrower modes.

On the basis of the above considerations one would expect that
rather large amounts of power would be needed to suppress a mode of
any strength.   This would be particularly true if Δ_{12} were large
which is desirable in many applications.   We find however, that when
I_1 is large it can still be suppressed with only a moderate amount of
power.   The I_1 vs I_2 dependence also deviates from a linear relationship.

A tentative explanation may be developed with the aid of the following argument.   Equation (3) represents a weak turbulence limit since
there is no provision for the reaction of the wave amplitude dependent
scattering on the spectral width.   We allow for this by modifying
Eq. (3);

$$\lambda(\omega) \simeq \frac{\alpha}{\pi} \sum_j \frac{I_j[\Gamma_j/2 + \lambda(\omega)]}{(\omega - \omega_{jo})^2 + [\frac{\Gamma_j}{2} + \lambda(\omega)]^2} \, . \qquad (11)$$

There will then be a strong turbulence limit for I_1 where

$$\lambda \simeq (\frac{\alpha}{\pi})^{1/2} I_1^{1/2} \quad , \tag{12}$$

a result not unlike that of Dupree (1967). Whether or not this limit is reached in a particular situation depends on whether or not the growth rate is greater than Γ. In the limit of Eq. (12) ($\lambda > \Gamma$, $\lambda < \Delta_{12}$) the dependence of I_1 upon I_2 becomes

$$I_1 \simeq 4 \frac{\pi \gamma^2}{\alpha} \frac{1}{\left[1 + \dfrac{\alpha I_2}{\Delta_{12}^2} \right]^2} \tag{13}$$

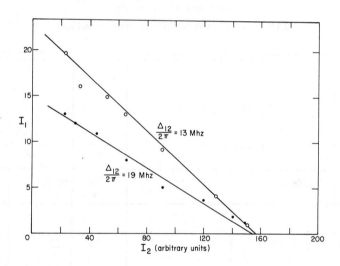

Fig. 4. Beam flux distributions for suppressed
and unsuppressed fluctuations.

A particular strong turbulence case is shown in Fig. 4. The shutting off of a strong natural oscillation by a higher-frequency, driven oscillation also has a pronounced effect on velocity space diffusion. Suppression of the natural turbulence decreases the beam width by a factor of 3, Fig. 5, although the net energy in the spectrum is slightly greater. The results of Figs. 3 and 4 are summarized in Table I.

In conclusion we find that the theory of Brownian motion when applied to electron plasma oscillations of moderate width is capable of

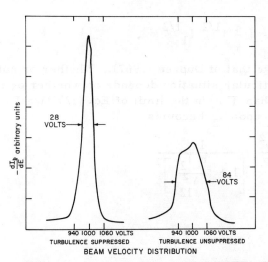

Fig. 5. Beam velocity distribution.

Fig.	I_1 at $I_2 = 0$ (ergs cm^{-3})		$\frac{\Delta_{12}}{2\pi}$	$\frac{4\,\Gamma_1\,\Gamma_2}{\Delta_{12}^2 + \Gamma_2^2}$	
	meas.	calc.	Mhz	meas.	calc.
3	6.0×10^{-5}	4.0×10^{-5}	13	.13	.15
3	4.3×10^{-5}	3.1×10^{-5}	19	.09	.07
4	1.67×10^{-3}	1.2×10^{-3}	14	--	--

Table I. Summary of experimental results.

explaining several nonlinear phenomena. The model is of some interest since its application is elementary and it is readily extended to take care of strong turbulence in a phenomenological way. Since the coupling does not depend upon whether or not the dispersion relation allows decay it is an elementary extension of the technique to use a high gain feedback loop rather than an external oscillator in order to accomplish stabilization. Further work on establishing the relationship of the present model to the more conventional approaches to nonlinear behavior is necessary.

ACKNOWLEDGMENTS

This work was supported in part by USAFOSR Contract No. AF 49(638)1481 and in part by a grant from the Research Corporation.

DYNAMIC-STABILIZATION EXPERIMENTS IN PINCHES

8.1 DYNAMIC STABILIZATION OF THE Z-PINCH[*]

J. A. Phillips, P. R. Forman, A. Haberstich, and H. Karr
Los Alamos Scientific Laboratory, Los Alamos, New Mexico 87544

ABSTRACT

Results of dynamic stabilization of the m = 1 kink instability of a low-current z-pinch ($\leq$ 40 kA) by (1) high-frequency linear-quadru-pole magnetic fields and (2) high frequency oscillating B_z magnetic fields are described. In method (1) the m = 1 instability of wavelengths of the order of the interelectrode spacing ($\sim$ 56 cm) are stabilized but the short wavelength instabilities of 5 to 12 cm are dominant and are not suppressed. The observed instabilities have a reduced growth rate with stabilization in agreement with MHD theory. The stabilizing field is also seen to (a) distort the cross section of the plasma column into an ellipse whose major axis increases with time and (b) to increase the total number of electrons in the discharge above the initial gas filling due to release of gas from the walls. MHD theory predicts growth times larger than the duration of the experiment for dynamic stabilization of an oscillating B_z magnetic field, method (2). Experimentally significant ($\sim$kA) j_θ discharge currents are driven parallel to the B_θ fields of the z-pinch near the walls by the stabilization fields. These j_θ currents shield out most of the time varying fields from the central region and prevent a test of the simple theory. Longitudinal limiters parallel the axis help to suppress these j_θ currents and stable times are increased by~75%.These experiments on dynamic stabilization of the z-pinch have been marginally effective in slowing down the growth rate of the m = 1 instability. Complete stability theoretically requires higher frequencies and currents than those used in the experiment but these would be technically difficult. However the data also show that as the dynamic field strengths and frequencies are raised there may be undesirable dangerous side effects: (1) the growth of driven oscillations and (2) the shielding out of the h.f. fields by induced currents in the residual plasma outside the pinched plasma column which may prevent dynamic stabilization from being effective.

---

[*]Work performed under the auspices of the U. S. Atomic Energy Commission.

I. INTRODUCTION

High-frequency stabilization of the magnetohydrodynamic instabilities characteristic of the z-pinch has been studied both theoretically and experimentally. We have been encouraged by earlier
results (Osovets and Sinitsyn, 1965; Bobyrev and Fedyanin, 1963)
to examine dynamic stabilization by two methods: 1) the oscillating
field of a longitudinal multipole and (2) an oscillating external B_z
field. Containment was provided by the self-magnetic field of the
pinch current and stability was maintained by the high frequency
magnetic fields. Stabilization of the m = 1 kink mode was investigated with the objective of determining the wavelength of the instabilities, growth rates, and confinement times. The theoretical considerations and experimental results for each method are described.

II. MAGNETIC QUADRUPOLE

The equation of motion[†] of the pinch discharge for the kink instability included the force developed by the interaction of the discharge current with the magnetic field of the longitudinal quadrupole.
If the oscillating quadrupole current has the form $I_Q \cos \omega t$, a
Mathieu equation results,

$$\frac{d^2\xi}{dz^2} + (\alpha - 2q \cos 2z)\xi = 0 \tag{1}$$

where ξ is the displacements amplitude, and $z = \omega t$. With $M = \pi a^2 \rho$, a is the plasma radius and ρ the mass density,

$$\alpha = \frac{-4I_z^2}{M\omega^2 c^2}\left(\frac{2\pi}{\lambda}\right)^2 \ln\frac{\lambda}{\pi a}, \text{ and } q = \frac{40}{M\omega^2 c^2 b^2} I_Q I_z.$$

I_z is the Z-current, λ the perturbation wavelength, I_Q the current
in each longitudinal conductor, and b is the radial conductor position.
Figure 1 shows the first stable region of the Mathieu equation. The
magnitude and frequency of the quadrupole current necessary for
stabilization are obtained from the bounds of the stable region. The
limits lead to the approximate stability conditions,

$$\frac{4\sqrt{2\pi}\,I_z}{\omega\lambda c}\sqrt{\frac{\ln\frac{\lambda}{\pi a}}{M}} < \frac{40\,I_Q I_z}{\omega^2 c^2 b^2 M} < 1. \tag{2}$$

[†] P.R. Forman, A. Haberstich, H. Karr, J. Phillips, and
A. Schofield, submitted to Physics of Fluids.

These conditions are similar to those of Osovets (1961) but differ by a numerical factor because of the inclusion of the plasma diamagnetic currents.

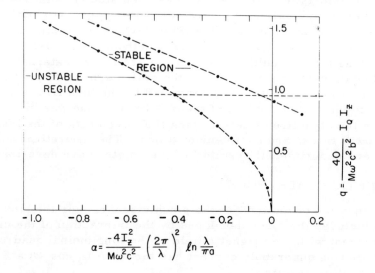

$$q = \frac{40}{M\omega^2 c^2 b^2} I_Q I_z$$

$$\alpha = \frac{-4 I_z^2}{M\omega^2 c^2} \left(\frac{2\pi}{\lambda} \right)^2 \ell n \frac{\lambda}{\pi a}$$

Fig. 1.   First stable region of the Mathieu equation.   The dashed horizontal line corresponds to the experimental parameters.

The dashed curve in Fig. 1 corresponds to the typical experimental values, I_z = 25 kA, I_Q = 13.7 kA, ω = 4.7×10^6/sec, b = 3.8 cm and M = 5×10^{-7} gm/cm.   From the intercepts of this line at α = -.425 and α = 0 with the boundaries of the stable region, the theory predicts stability for wavelengths $\lambda > 20$ cm.   Other experiments (Osovets and Sinitsyn, 1965) have been encouraging in that it was postulated that if the long wavelengths appear first and are stabilized the shorter ones do not become important.

The growth rates of the short wavelengths in the unstable regions of the Mathieu diagram were calculated from the exponential factor of the Floquet solutions of the Mathieu equation for our experimental conditions.   Figure 2 is a plot of the kink growth rates with dynamic stabilization shown as a solid line and without as a dashed line.   The short wavelengths, 5 to 20 cm, remain unstable but their growth rates are reduced.

A schematic drawing of multipole conductors and circuit diagram used is shown in Fig. 3.

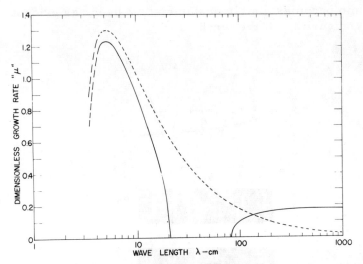

Fig. 2. Growth rates of the kink instability as a function of wavelength with (solid line) and without (dashed line) stabilization.

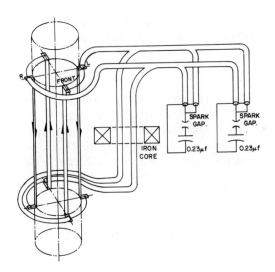

Fig. 3. Schematic drawing of multipole conductors and circuit diagram.

A. Results

1. Figure 4 shows photographs of the plasma column taken at selected times in the discharge cycle. Each exposure shows two side views of the pinch. Without stabilization the 1-cm-radius

plasma column becomes visibly unstable at ~2 μsec, as evident in
the kinking and curling of the pinch.

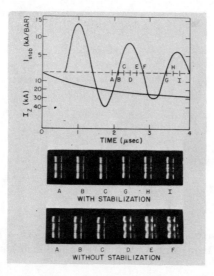

Fig. 4.   Photographs of the pinch discharge with and without
stabilization taken at selected times in the discharge cycle.

The m = 1 instability wavelengths vary over the range ~5 to 12 cm
with the amplitude growing in time until the discharge reaches the
tube walls at ~2.7 μsec.   With stabilization the growth rates are
noticeably reduced, ~30%.   The wavelengths ~5 to 12 cm are still
predominant and lie outside the theoretically stabilized region in the
Mathieu diagram in Fig. 1, but are within the range of the computed
growth rates of Fig. 2.

    2.   Comparison of the damping of the stabilization current wave-
form with and without plasma shows a substantial transfer of energy
to the discharge volume.   With the z-pinch ~17 joules are lost by the
stabilizing circuit by the time of the first current peak at 0.34 μsec.
We believe that this energy loss is due to an electrical breakdown
occurring along the inside of the discharge tube wall with the wall
currents returning along the axis of the discharge.   The induced
electric field along the wall is ~220 V/cm.   This wall breakdown
should release gas from the walls which could be the source of
additional plasma seen by interferometry.

    3.   The onset of the kink instability can also be determined by
flux loops around the discharge tube which detect axial magnetic

field components. The measurements show that stabilization delays the onset in agreement with image converter photographs and with theory. Also, increases in the stabilization field further decreased the stability growth rate as predicted but, as seen below, further increases above a critical magnitude would not be advantageous.

4. Interferometry determines (a) the radial electron density distribution, (b) distortions of the pinch by the stabilization field, and (c) the total number of electrons in the discharge as a function of time. Figure 5 shows density contours with the stabilization field.

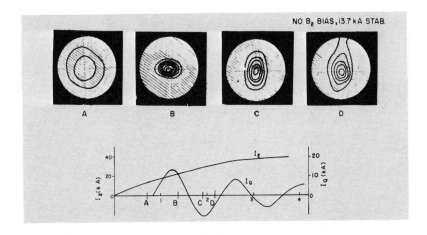

Fig. 5. Density contours obtained from interferograms.

Note the elliptical shape with the ellipse major axis shifting $90°$ when the conductor current is reversed. In Fig. 5d the density contours extend to the wall as a result of overdriving the plasma with the stabilizing field. The total plasma content, determined by integration over the interferograms, is found to equal to within $\sim 15\%$ the initial gas filling for the first $\sim 2\ \mu sec$. After $2\ \mu sec$ the total number of electrons began to rise at a rate which increases with the magnitude of the stabilization currents. This rise is believed to be caused by the contact of the plasma column with the walls and by the wall currents discussed above.

The eccentricity of the discharge cross section was compared with the theoretical prediction (Ribe, 1969) by letting

$$r = a(1 - \delta_2 \cos 2\theta), \tag{3}$$

$$\text{and} \quad \delta_2 = 8\left(\frac{a}{b}\right)^2 \frac{I_Q}{I_z}. \tag{4}$$

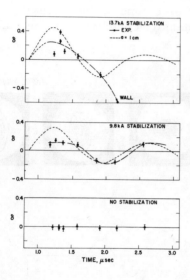

Fig. 6. Elliptical distortion of the plasma cross section due to the quadrupole stabilizing field.

The results are shown in Fig. 6. Without stabilization the discharge was symmetric within experimental error and $\delta_2 = 0$. With a conductor current of 9.8 kA, $\delta_2 = +0.12$ on the first half cycle, -0.15 on the second, and $+0.10$ on the third. With the larger dynamic stabilization amplitude, δ_2 became larger until the plasma reaches the wall in the second half cycle. This effect limits the magnitude of the stabilization field that can be used. The theoretical value of δ_2 is shown as a dotted line in Fig. 6.

B. Conclusions

The experimental results show that the shorter wavelengths ($\lambda \lesssim 20$ cm) are the most dangerous and their growth rates are only reduced in agreement with theory. Stabilization of these shorter wavelengths requires very high frequencies and high stabilizing currents.

Several undesirable effects are observed (a) a significant energy transfer from the stabilizing circuit into the residual gas outside the pinch column, (b) an increased line density of electrons in the tube with increasing stabilization fields and (c) at the higher stabilizing fields the cross section exhibits oscillations which grow rapidly until the plasma reaches the tube wall.

III. OSCILLATING B_z FIELD

The stability conditions are given by Bobyrev and Fedyanin (1963) from the equation of motion of a flexible conductor bending along a line of magnetic force in the limit of small amplitude perturbations. With the application of a high frequency $\hat{B}_z$ field the equation of motion assumes the form of the Hill equation. From the bounds of the stable region the necessary limit of frequency and amplitude of the stabilizing field are given by

$$\omega^2 > \frac{B_\theta^{\,2}(a)}{M} \quad,$$

$$\tilde{B}_z \geq B_\theta(a)$$

where $B_\theta(a)$ is the pinch field at the plasma surface.

The growth rates have been calculated[†] for the conditions under which the theory does not indicate complete stability. The model used is a sharp boundary incompressible plasma column which obeys the ideal MHD equations. Again the equation of motion is the Hill equation and growth rates are obtained from the Floquet solutions. It is found that for the experimental parameters, $I_z = 25$ kA, $\omega = 4.7 \times 10^6$/sec, $M = 5 \times 10^{-7}$ gm/cm and $\tilde{B}_z \gtrsim 3$ kG, the growth rates are too small to be observed, i.e., there should be stability for the duration of the experiment.

A. Results

Photographs of the z-pinch taken at selected times in the discharge cycle and flux loop measurements show an increase in the stability of the z-pinch with the modulating $\tilde{B}_z$ field with the onset of the kink delayed by 25%. Strong luminosity associated with a

---

[†] P.R. Forman, H. Karr, and J. Phillips, to be published.

discharge in the residual plasma near the tube walls, however, was observed outside the pinch column.

Flux loops around the discharge tube showed that the $\hat{B}_z$ penetrated into the discharge on the first quarter cycle but a considerable fraction was shielded on the second quarter cycle. Radial B_z and B_θ magnetic field distributions, Fig. 7, also showed this behavior.

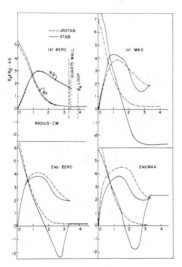

Fig. 7. Radial B_z and B_θ magnetic field distributions taken at times of successive $\hat{B}_z$ zeros and maxima.

Attempts to decrease this field effect by changing gas pressure, tube wall material, conditioning of the wall, and timing of the stabilization field relative to the pinch discharge had no significant effect. It was postulated that as the B_z field decreased to zero, plasma on magnetic field lines moved radially outward and bombarded the tube walls making the region electrically conducting. An I_Θ current was then induced near the wall.

In a modified pinch discharge tube the I_Θ currents were interrupted by placing two diametrically opposite longitudinal limiters inside the tube over the full length of the discharge tube, Fig. 8. The limiters were made of Pyrex and extended radially inward from the wall to within a distance of 1 cm from the plasma column, radius ~1 cm. Flux loop measurements, Fig. 9, and probe distributions, Fig. 10, showed that in the radial region between the inside edge of the limiters and the tube wall there was the free flow of B_z

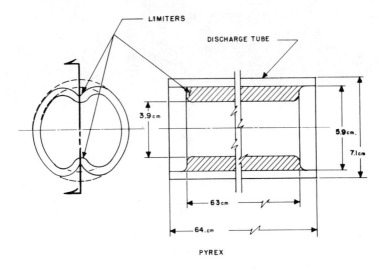

Fig. 8.   Location of longitudinal limiters in the discharge tube.

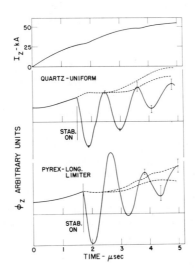

Fig. 9.   Flux loop signals as functions of time for a uniform
quartz tube and for a Pyrex tube with longitudinal limiters.

flux.   Mach-Zehnder interferometry also showed wall impurities to
be significantly reduced.   Allowing the B_z field to penetrate close to
the pinch column by the limiters further delayed the onset of the
kink stability.

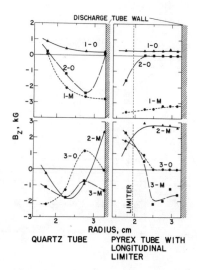

Fig. 10.   Radial B_z magnetic field distribution for two tubes, (a) uniform quartz tube and (b) Pyrex tube with longitudinal limiters at times of successive $\dot{B}_z$ zeros and maxima.

The signals from flux loops indicated that the average time of onset was delayed by ∽75%, from ∽3.1 μsec without the stabilization field to ∽5.4 μsec with stabilization. At these late times in the discharge cycle, (a) the peak B_z had decayed to ∽30% of that on the first half cle cycle by the damped LC driving circuit, (b) loss of plasma out the ends decreased the line mass of the pinch column and (c) the I_z current had increased to ∽40 kA. All these factors result in less effective stabilization by the B_z. The increase in stable times as shown in this experiment may not be the maximum that can be obtained by this method. An extension of the experimental parameters would be necessary to show this.

IV. GENERAL REMARKS

Stabilization of a wide range of frequencies of the kink mode of the z-pinch by one stabilization frequency appears difficult. In the z-pinch the fastest growing wavelengths are the shortest and stabilization of these wavelengths requires high frequencies and field magnitudes estimated to be an order of magnitude above those used in the present experiment. Such frequencies are technically difficult and wall breakdown would be expected to become more severe due to the higher electric fields. Also an increase in the amplitude of the stabilizing fields may cause excessive plasma distortions.

The modulating B_z field experiment has clearly demonstrated the importance of conducting plasma outside the pinch column. Wall currents may be induced which exclude effectively further changes in field strengths. It has been found, however, that limiters judiciously placed may interrupt those currents and permit the flow of magnetic flux.

8.2 DYNAMIC STABILIZATION EXPERIMENT ON A LINEAR SCREW PINCH

G. Becker, O. Gruber, H. Herold

Institut für Plasmaphysik, Garching, Germany

ABSTRACT

In a linear screw pinch (T_i = 50 to 350 eV; β = .5 to .7) helical m = 1 instabilities with growth rates ω_i of about .3 to 3 $\times$ 10^6 sec^{-1} were observed. A modulation of the quasi-stationary B_{zo} field component with amplitude ratios $\mathcal{E}$ = B_z / B_{zo} up to .2 and with frequencies ω_s of 7 and 10 $\times$ 10^6 sec^{-1} was applied for dynamic stabilization. A reduction of the growth of the m = 1 mode could be achieved provided that $\mathcal{E}\omega_s$ was higher than a critical value for a given ω_i. Dynamic shear of the field lines in the plasma boundary and inertial effects appeared during the enforced radial oscillation of the plasma surface. An upper limit for $\mathcal{E}\omega_s$ beyond which breakdown close to the wall occurred was observed.

I. INTRODUCTION

Experiments on high-β linear and toroidal screw pinches (Gruber, Herold, Wilhelm, and Zwicker, 1969; Little, Newton, Quinn, and Ribe, 1968) have shown that they are MHD unstable. The experimentally observed growth rates ω_i of the prominent helical m = 1 modes agree with those derived from constant pitch theory (Schuurmann, Bobeldijk, de Vries, 1969) or diffuse pinch theory (Friedberg, 1969). Since the values of ω_i are relatively low, at least compared to the results of sharp-boundary theory for the same equilibrium position, a dynamic stabilization of the m = 1 mode may be feasible in terms of the stabilization frequency needed ($\omega_s > \omega_i$). In a first approach an oscillating B_z-field was superposed on the basic longitudinal field component B_{zo} of the screw pinch (B_z = B_{zo} + $\tilde{B}_z \sin \omega_s t$). According to Wobig and Tasso(1970), inertial forces are considered to be effective for stabilization and a stabilization condition $1 > \Delta r_p / r_{po} > C \omega_i / \omega_s$ may be relevant (r_p = plasma radius). It may be possible to work in resonance with natural oscillations of the plasma column (m = 0 and k = 0, in our case) to save evergy of the driving h.f. field, provided that $\omega_s \sim 2\omega_o/n > \omega_i$.

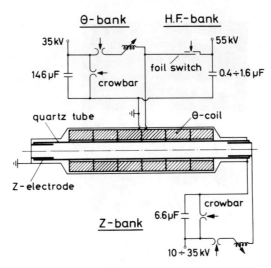

Fig. 1.   Linear screw pinch with basic circuits.

II.   EXPERIMENTAL SET-UP AND PLASMA PROPERTIES

The screw pinch experiment with the basic circuits to produce B_{zo} (up to 30 kG) and I_{zo} (10 to 90 kA) is sketched in Fig. 1.   The circuits for preionization and bias fields are omitted.   The rise-time $(T/4 \lesssim 4 \; \mu sec)$ of the I_{zo} and B_{zo} are matched by variable inductances.   The coil length and the electrode distance   L   are 100 cm (I. D. 12 cm) and 120 cm respectively.   The oscillating current $\tilde{I}_\theta$ produced by a bank of high Q capacitors   (Q = 90 at f = 2Mc/sec) is fed into the B_z coil via a foil switch with low in-ductance (2 nHy), low resistance $(< 1 \; m\Omega$ at $\tilde{I}_\theta > 200$ kA) and low jitter (10 nsec).   Mostly, a $\tilde{B}_z$ with $\omega_s/2\pi = 1.1 \times 10^6 \; sec^{-1}$ and amplitudes up to 3 kG was applied.   For some experiments $\omega_s/2\pi$ was increased to $1.7 \times 10^6 \; sec^{-1}$ with $\tilde{B}_z = 1.8$ kG.   The e-folding time of the oscillation amplitude with plasma were typically about 8 μsec for the lower frequency.   The plasma parameters in the screw pinch $(\tilde{B}_z = 0)$ were: T_i = 50 to 350 eV; $n_e = 2$ to $10 \times 10^{16} cm^{-3}$. The growth rates $\omega_i/2\pi$ of the helical m = 1 instability varied between .5 and $5 \times 10^5 \; sec^{-1}$ for I_{zo} from 20 to 90 kA.   So the condition $\omega_s > \omega_i$ was well fulfilled.

III.   RESULTS

To test the excitation of otherwise stable modes by parametric resonances the oscillating B_z was applied to the pure θ-pinch $(I_{zo} = 0)$.   Two different ω_s were used and the delay time for the h.f. bank (Δt_{HF}, see Fig. 5) was varied with $\mathcal{E} = \tilde{B}_z / B_{zo}$ up to .2.

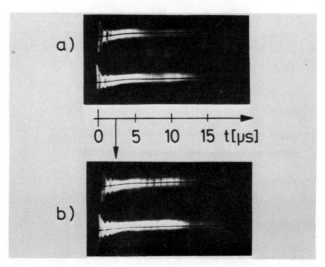

Fig. 2. Stereoscopic streak pictures of a θ-pinch plasma without (a) and with (b) superposed oscillating field.

There was in no case deterioration of the θ-pinch stability. Figure 2 gives an example of the stereoscopic side-on streak pictures used to observe the stability behaviour. The oscillatory motion of the plasma radius was analyzed by shot-to-shot density measurements by means of holographic interferometry and by time-resolved measurements using a laser interferometer side-on. The magnetic fields were measured with radially inserted magnetic probes. It turned out that the radial surface motion corresponds to $|\Delta r_p|/r_{po} \approx |\tilde{B}_z|/B_{zo}$ rather than $|\Delta r_p|/r_{po} \approx \frac{1}{2}(|\tilde{B}_z|/B_{zo})$ as expected for adiabatic motion with $\omega_s/2\pi \ll v_a/r_p$ (r_p is the radius where $n(r)$ has dropped to $(1/e) n_{max}$ and v_a is the Alfven velocity). There was also a phase shift of 120° to 150° between the forced oscillation of the plasma and the driving field. Figure 3 shows density profiles taken at extreme positions of r_p. These results indicate an inertial behaviour of the plasma motion. ω_s/ω_o was between .2 and .5. These measurements are also representative for the screw pinch since $B_{\theta o}^2 \ll B_{zo}^2$. When B_z is applied to the screw pinch an appreciable decrease of $m = 1$ instability growth can be achieved, leading to a delay $\Delta t_c = \tilde{t}_c - t_c$ of wall contact. Here t_c and $\tilde{t}_c$ are the times between the start of the main discharge and wall contact without and with dynamic stabilization respectively. This is shown in stereoscopic streak pictures taken with 40 and 80 mtorr deuterium in Fig. 4. Figure 5 gives a typical example for the currents and fields. In order to get an observable stabilizing effect, $\mathcal{E}$ has to be greater than a critical value which depends on ω_i, provided that $\omega_s > \omega_i$; Table 1 gives examples for $\omega_s / 2\pi = 1.1$ MHz/sec. The critical $\mathcal{E}$ is lower if

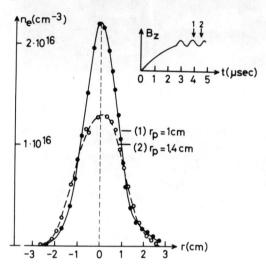

Fig. 3. Density profiles for extreme positions of the plasma radius taken at times indicated on the upper right. (p_0 = 20 mtorr, deuterium; $\mathcal{E}$ = .15)

ω_i, sec^{-1}	$\mathcal{E}_{crit}$	t_c, μsec	$\tilde{t}_c$, μsec
0.5×10^6	0.1 - 0.12	10	15
0.75×10^6	0.18 - 0.2	7	12

Table 1

ω_s is increased so that $\mathcal{E}\omega_s$ is approximately constant for a given ω_i, in agreement with $\Delta r_p/r_{po} > \omega_i C/\omega_s$. But the experimental material is not yet sufficient to confirm this condition. With dynamic stabilization the residual growth of the m = 1 instabilities is no longer exponential but rather of an irregular character. Also the appearance of the m = 2 mode cannot be excluded in some cases. The growth of the m = 1 instability can still be reduced when the stabilizing field is switched on after an observable unstable displacement ξ ($\xi \lesssim r_p$) has taken place. This gives some hope for application of feedback schemes. Probe measurements reveal that there is also a dynamic shear of the field lines in the plasma surface during oscillations of r_p which may have a stabilizing effect. Figure 6 shows the field components B_z and B_θ for two successive extreme positions of r_p. The broad radial distributions of I_{zo} stays constant during the motion of the dense plasma column.

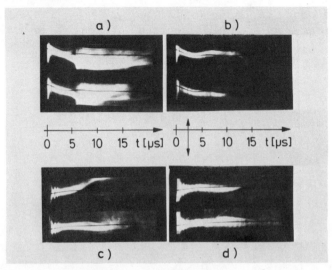

Fig. 4. Stereoscopic streak pictures of screw pinches without (a, c) and with (b, d) dynamic stabilization. (a, b, 40 mtorr, D_2, c, d 80 mtorr, D_2).

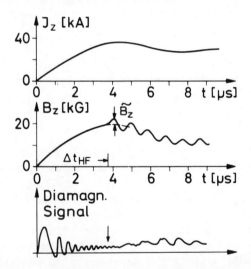

Fig. 5. Longitudinal current, external B_z-field and diamagnetic signal of a screw pinch with dynamic stabilization.

The oscillation of B_z together with the inertially caused over-shoot of the plasma radius give a variation of the pitch of the field lines $\mu_{(r_p)} = (B_\theta / r B_z)_{(r_p)}$ of about 5 % between the two extreme positions of r_p for $\tilde{\mathcal{E}} \approx 0.1$.

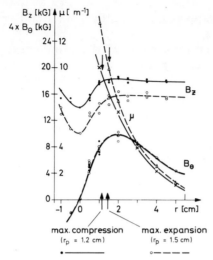

Fig. 6.   Radial field profiles and pitch of field lines
$\mu = B_\theta / r B_z$ taken at extreme positions of the plasma radius during
dynamic stabilization (p_0 = 40 mtorr, deuterium; $\mathcal{E}$ = 0.1; I_z = 30kA).

The efficiency of the stabilization is found to depend on the
switch-on time of $\tilde{B}_z$ relative to the phase of the natural $m = 0$
plasma oscillation.   More effective stabilization occurs when the
first maximum of $\tilde{B}_z \sin \omega_s t$ coincides with a minimum of r_p.   This
behaviour may have to do with transients and damping of the natural
oscillations.   At high stabilization frequencies and high field ratios,
breakdown close to the wall occurred which caused a strong damping
of the oscillatory current.   The breakdown condition
$r_g \mathcal{E} \omega_s / 2 \pi \geq 1.2$ (r_g = wall radius in cm, ω_s in MHz/sec) given
by Chodura (1964) holds almost exactly.

IV. CONCLUSIONS

The experiments show that an improvement of plasma stability
in a screw pinch by enforced oscillation of the plasma surface is
possible.   The stabilization may be achieved by the combined effect
of inertial forces and dynamic shear in the diffuse plasma column.
Observed resonance-like phenomena as phase shift and resonance
elevation of the amplitude of r_p without excitation of new instabilities,
suggest the potential advantage of working in resonance with
natural plasma oscillations.   The method of enforced plasma oscillation, applied at higher plasma temperatures, is questionable.
ω_i and ω_o scale approximately with $\sqrt{kT}$ and so the product

$\omega_s \mathcal{E}$ necessary may be beyond the breakdown limit or beyond
technological feasibility.   Also, damping factors at high temperatures,
important for resonance elevation and stability, are hard to predict.

This work was performed as part of the joint research program
between the Institut für Plasmaphysik, Garching and Euratom.

DYNAMIC-STABILIZATION EXPERIMENTS IN FLUIDS AND SOLIDS.
CUSP LOSS CONTROL

9.1 DYNAMIC STABILIZATION OF HYDRODYNAMIC INTERCHANGE
INSTABILITIES - A MODEL FOR PLASMA PHYSICS

G.-H. Wolf
Institut für Plasmaphysik, 8046 Garching near Munich
Federal Republic of Germany*

ABSTRACT

Dynamic stabilization and parametric resonances of superposed
fluids with sharp and diffuse interfaces are described. An analogous
method is used to derive the stabilizing potential for the short-wavelength flutes for the standing-wave scheme of a high-β plasma
column.

I. INTRODUCTION

Dynamic stabilization has attracted interest in connection with
the problem of plasma instabilities, particularly MHD instabilities
in high-β configurations. Most previous experimental attempts to
stabilize such plasma instabilities have, however, been rather
limited by the short duration of the plasma state of interest, this
being due, for instance, to the fast decay of the applied oscillating
fields. Thus, although experimentally the effect of dynamic stabilization on high-β plasma confinement has been shown, the resulting
prolongation of the confinement was only of the order of a few
growth times of the relevant MHD instabilities.

An experimental example of dynamic stabilization on a long
time scale, on the other hand, is afforded by the pendulum with
oscillating support, which shows only a single eigenfrequency. This
impairs the compatibility of the pendulum experiment with the
situation in plasma physics, where generally a whole spectrum of
possible modes (stable or unstable) has to be considered. However,

---

*Work performed in association with Euratom.

applying dynamic stabilization to equilibria of fluids allows us not
only to conduct steady-state experiments with easy access to the
various parameters of interest, but also to study problems similar
to those of magnetically confined plasmas.

This paper deals with the stabilization of the hydrostatic interchange instability, i.e. the Rayleigh-Taylor instability (Lord
Rayleigh, 1900), of both sharp and diffuse boundaries, particularly
in connection with the question of parametric resonances (Meixner
and Schäfke, 1954). Furthermore, and as a link with plasma
problems, the original method of deriving the Rayleigh-Taylor
stabilization condition (Wolf, 1969a) is applied to obtain the stabilizing
potential - with respect to flute perturbations of short azimuthal
wavelengths - of the standing wave scheme (Berge, 1969a; 1969c;
Bodin, Butt, McCartan, and Wolf, 1969) of a high-β plasma column.

II. STABILITY AND PARAMETRIC RESONANCES

The unstable gravitational equilibrium of superposed liquids can
be dynamically stabilized by oscillating the liquids in the vertical
direction (Wolf, 1969a). Originally, the stability condition was
derived (Wolf, 1969a) from the time-averaged stabilizing potential
U_{eff} of a single particle of mass M subject to a harmonic force f
($\overline{f^2}$ is the time average) of frequency ω with $\omega \gg \Omega_K$, where Ω_K
are the possible eigenfrequencies of the static system. Then
(Landau and Lifshitz, 1962)

$$U_{eff} = U_s + \overline{f}^2/2\,M\,\omega \quad . \tag{1}$$

U_s is the potential of the static case. This yielded the stability
condition (Wolf, 1969a)

$$b_m/g \; \gtrsim \; \sqrt{2}\,\omega/\Omega_K \tag{2}$$

where b_m is the maximum instantaneous acceleration due to the
enforced harmonic oscillation and g is the acceleration of gravity.
The treatment of dynamic problems via U_{eff} also allows us to
derive stable dynamic equilibria (Wolf, 1969a, Wolf and Berge, 1969).
However, the access to the problem of parametric resonances is
more readily gained by using Mathieu's stability chart (Meixner and
Schäfke, 1954; Kotowski, 1943). If damping effects are neglected
the equation of motion describing the time development of a small
perturbation ξ of the equilibrium position is of the Mathieu type:

$$\ddot{\xi} + (1 + \frac{b_m}{g} \cos \omega t)\, \Omega_K^2\, \xi = 0 \tag{3}$$

If furthermore only the plane with imaginary values of Ω_K is considered and interest is concentrated on the main stability region, the lower limit of this region is fairly well described by Eq. (2), while in the parameter regime investigated experimentally the upper limit is approximated by:

$$b_m/g \approx \alpha(\omega/\Omega_K)^2 \tag{4}$$

with $\alpha \approx 0.45$ for zero damping $(\delta^2 = 0)$ and, in the presence of an additional damping term introduced in Eq. (3), with $\alpha \approx 0.7$ for $\delta^2 = 0.1$, where δ is the logarithmic decrement of the Ω_K (Wolf and Berge, 1969). Achieving dynamic stabilization thus implies that Eqs. (2) and (4) have to be satisfied simultaneously. Consequently, for a given oscillator power the possible values of Ω_K have to show not only a lower limit (in the case of sharp boundary layers determined by the vessel dimensions) but also an upper one.

In the experiments with sharp boundary described below the upper limit is given by viscosity damping. Violating Eq. (4) then excites "parametric resonances" which occur at that wavelength L_p where the counteracting influence of damping (increasingly effective for short wavelength modes) balances the increase in Ω_K (proportional to L_K^{-1} without damping). Measureing $L_p(\omega)$ allows us to determine the corresponding undamped eigenfrequency $\Omega_p(\omega)$ e.g. by using Eq. (1) or (9) of Wolf (1969a), which is required to enter the Mathieu chart.

III. SHARP-BOUNDARY EXPERIMENTS

Experiments have been carried out for the fluid combinations of: (1) viscous liquids (e.g. liquid paraffin) in the viscosity range of 1-6 poise against air (Wolf, 1970a) and (2) mercury against motor oil of 12 poise. Figures 1-3 show typical photographs (for details see figure captions) of the liquid paraffin-air interface in parameter regimes where either the lower or the upper demarcation of the stability region is approached or even crossed. Figure 3, for instance, gives an example of some excited wavelengths $L_p(\omega)$. Each of the corresponding values $2(b_m/g)(\Omega_p/\omega)^2$ represents the empirically found lowest point (x) of the upper stability boundary (for a certain Ω_p) in Mathieu's stability chart, so that the whole set of $\Omega_p(\omega)$ allows us to obtain the envelope of those boundary curves. This envelope is then the upper stability limit for the whole system including all the possible modes Ω_K.

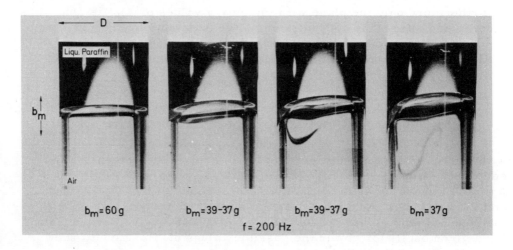

Fig. 1. Photographs of liquid paraffin ($\eta \approx 0.96$ poise) super-imposed upon air. The value of $b_m \approx 37$ g is just below the lower stability limit (Eq. (2)). The inner vessel diameter is D=1.6cm, f is $\omega/2\pi$.

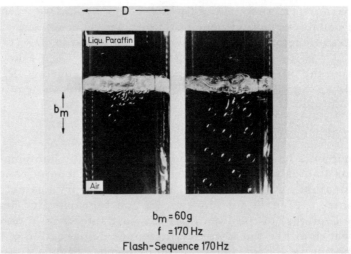

Fig. 2. Photographs corresponding to those in Fig. 1, but with b_m slightly beyond the upper stability limit (Eq. (4)), thus showing rain-like falling drops.

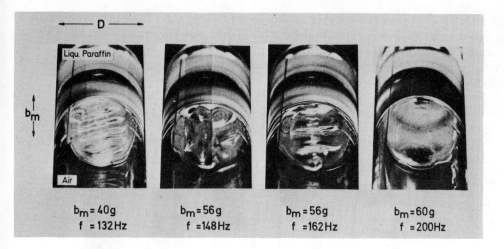

Fig. 3. Photographs corresponding to those in Figs. 1 and 2, but with a different angle of view and with various combinations of b_m and f just approaching the upper stability limit (Eq. (4)), which show parametrically excited wavelengths $L_p (\omega)$.

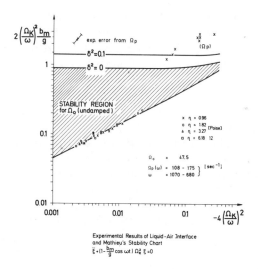

Fig. 4. Results of the liquid-air experiments plotted in a Mathieu chart which shows theoretical upper limits for both zero damping ($\delta^2 = 0$) and weak damping (Kotowski, 1943) ($\delta^2 = .1$), where δ is the logarithmic decrement of the modes considered.

In Fig. 4 the experimentally found values $2(b_m/g)(\Omega_p/\omega)^2$ are plotted in a Mathieu chart together with the experimental results for the lower stability boundary. The experimentally found parametric resonances indicate that the effective damping is typically of the order $\delta^2 \approx 0.1$.

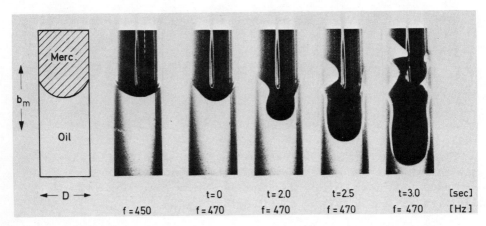

Fig. 5. Photographs of mercury superimposed upon motor oil ($\eta \approx 12$ poise) for $b_m = 55$ g and two values of $f = \omega/2\pi$, one just above and the other just below the lower stability limit (Eq. (2)). The time development of the resulting instability is shown. $D = 0.78$ cm.

Similar results are shown in Fig. 5 - 7 for the mercury-oil interface. Note, however, that here the value of m_o (see Eq. (9) of Wolf (1969a)) determining the lowest possible eigenfrequency Ω_K is not $3.68/D$ but $7.66/D$ because of the meniscus shaped mercury surface (Fig. 5) due to surface and interfacial tension. Furthermore, in order to obtain Ω_p from Fig. 6, the value of m_o is $10.66/D$.

Plotting these experimental results in a b_m/g versus ω plane rather than on a Mathieu chart gives empirical formulas for the "parametric resonance" stability boundary, which is independent of a knowledge of the $L_p(\omega)$ values:

$$\frac{b_m}{g} \approx 0.50 \frac{\omega}{2\pi} - 30 \qquad (5)$$

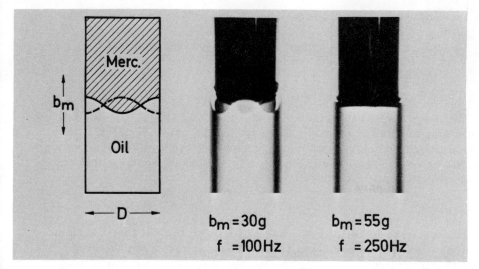

Fig. 6.   Photographs corresponding to those in Fig. 5 for two combinations of b_m and f. The left picture shows an example of a parametrically excited mode, the two extreme positions of which, photographed stroboscopically, are superposed. The right picture shows the interface in a regime where even the typical meniscus shape is suppressed by the stabilizing force.

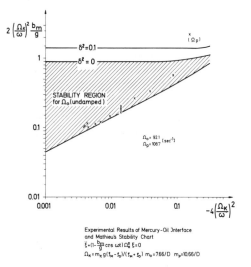

Fig. 7.   Results of the mercury-oil experiments plotted in a Mathieu chart corresponding to Fig. 4.

for the liquid paraffin-air interface (Wolf, 1970a) and

$$\frac{b_m}{g} \approx 0.43 \; \frac{\omega}{2\pi} - 9 \tag{6}$$

for the mercury-oil interface. Combining Eq. (5) or (6) with Eq. (4) would suggest that for large values of ω the excited $\Omega_p(\omega)$ are approximately proportional to $\sqrt{\omega}$; however, the rather large experimental error there did not allow this trend to be proved within the measured range of $L_p(\omega)$.

IV. DIFFUSE-BOUNDARY EXPERIMENTS

It has already been shown by Lord Rayleigh(1900) that a diffuse boundary between superposed fluids (with exponentially varying density) the eigenfrequencies Ω_K do not have a lower limit, but can approach zero with decreasing wavelength in the vertical direction. Consequently complete stabilization cannot be achieved, as shown theoretically by Wesson*. On the other hand the Ω_K do show an upper limit and parametric resonances can be suppressed even for ideal fluids.

To observe the influence of those modes $\Omega_K < \Omega_1$, where Ω_1 is the lowest frequency mode that can be stabilized due to Eq. (2), experiments were conducted using miscible liquids of different density. These were: a) an aqueous solution of ZnJ of density $\rho_h = 1.8 \text{ g/cm}^3$ with water superimposed and b) an aqueous solution of ZnJ diluted with glycerine of density $\rho_h = 1.8 \text{ g/cm}^3$ with a superimposed mixture of water and glycerine (density $\rho_i = 1.2 \text{ g/cm}^3$) such that the viscosity of both liquids was about 50 times that of water. While being oscillated the vessel containing the liquids was turned upside down. Typical photographs are shown in Fig. 8, where it can be seen that in the grossly stabilized state thin layers of relatively low and high density were continuously moving upwards or downwards respectively, as if being peeled off from the diffuse interface. This caused a gradual reduction in the density difference between the two liquids which would have led to a breakdown of the grossly stabilized state (in our case at $\Delta\rho \approx 0.13$, occurring after about 15-20 minutes) if the vibrator had not been always switched off after about 8 minutes due to overheating. The density profiles were obtained from light absorption measured

---

*   J. Wesson, to be published in Physics of Fluids.

photometrically.   The time development of these density profiles
is indicated in Fig. 9.   There the most obvious feature is that the

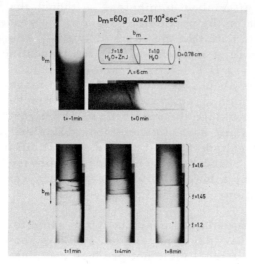

Fig. 8.   Typical photographs of superimposed miscible liquids
of different density; for t > 0 the heavier fluid is on top.   The
wedge filter was used for calibration.

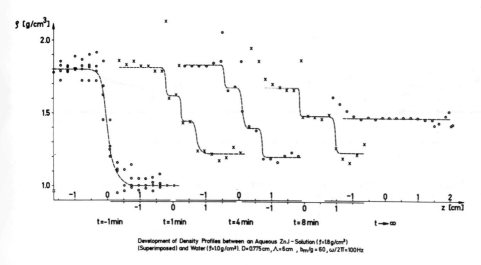

Fig. 9.   Density profiles taken from the photographs shown in
Fig. 8 by using light absorption.   The isolated experimental points
outside the dashed lines are ascribed to light refraction from the
strong density gradients.

boundary sheath originally showing continuous decay in density was transformed into three almost equally spaced layers, the upper one of which broke down after about 6 minutes. Although the decay time, τ_d, of $\Delta\rho/\rho$ depends on the particular length of the vessel, Λ, it might be of interest to note that the value of τ_d/τ_D, where τ_D is the instability growth time of the mode with the longest horizontal wavelength, λ_D ($\lambda_D = \pi D/1.84$), is of the order of $10^4 - 10^5$. On the other hand, τ_d is about two to three orders of magnitude shorter than that time after which an equivalent state was reached by normal diffusion. Roughly speaking, the unstabilized short vertical wavelength modes caused some sort of "anomalous diffusion" about two to three orders of magnitude faster than the normal diffusion process. However, parametric resonances were indeed suppressed; their only effect was to determine a lower limit of the boundary sheath thickness such that maximum $\Omega_K \lesseqgtr \Omega_p$.

The above results remained essentially the same for the case b) of the diffuse boundary layer experiments. This shows that the introduction of strong viscosity did not produce complete stabilization either, agreeing with a recent theoretical treatment by Wobig[*].

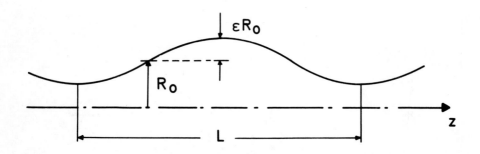

Fig. 10. Geometric parameters of the standing wave configuration ($R = R_o$ [$1 + \mathcal{E} \sin 2\pi z/L \cos \omega t$]) superimposed upon a square profile high-β plasma column.

[*]H. Wobig, private communication.

V. THE STANDING WAVE SCHEME FOR FLUTES OF LARGE AZIMUTHAL WAVE NUMBERS ($m \gg 1$)

Superimposing a standing-wave-like oscillation of frequency ω upon a square profile high-β plasma column (for geometric parameters see Fig. 10) results in a stabilizing effect for flutes $m \geq 1$, as derived by Berge (1969a, 1969c). U_{eff} ($m \gg 1$) can also be estimated from Eq. (1) in a simplified but less rigorous way. The maximum value of the harmonic force, f_m ($f^{-2} = f_m^2/2$), is

$$f_m = M \xi \, \Omega^2 \tag{7}$$

where Ω^2 are the eigenfrequencies (squared) typical of the oscillated system. They are composed of two terms, $\Omega_1^2 + \Omega_2^2$, the first of which

$$\Omega_1^2 = 2 \pi \, b_m / \lambda = m b_m / R_o \tag{8}$$

is due to the Rayleigh-Taylor flutes of wavelength $\lambda = 2 \pi R_o / m$, b_m is $R_o \, \bar{\mathcal{E}} \, \omega^2$, where $\bar{\mathcal{E}}$ is the axial average 2 $\mathcal{E}/\pi$. The second term, Ω_2^2, stems from the maximum instantaneous growth rate of a corrugated plasma column. In evaluating Ω_2 one has to take into account that the oscillation time $2\pi/\omega$ is postulated to be shorter than the propagation time of the perturbation along one corrugation period. Consequently, Ω_2 cannot be the Haas-Wesson growth rate, but that growth rate in which only the localized regions of bad curvature are counted. This leads to the estimate:

$$\Omega_2^2 = 16 \, c_s^2 \, m \pi \, \mathcal{E}/\gamma \, L^2 \tag{9}$$

where C_s is the sound speed and γ the adiabatic exponent. Combining Eqs. (1), (7), (8) and (9) yields

$$U_{eff} = M \, m^2 \, \mathcal{E}^2 \xi^2 \left\{ \frac{16 \pi C_s^2}{\gamma L^2 \omega} + \frac{2\omega}{\pi} \right\}^2 \tag{10}$$

By neglecting terms with ω^{-2} and inserting for M the line mass density $\pi R_o^2 \, \rho_o$ with ρ_o the local mass density, one obtains

$$U_{eff} = \xi^2 m^2 \varepsilon^2 \left(\frac{2\pi R_o}{L}\right)^2 \frac{4C_s^2 \rho_o}{\gamma \pi} \left\{1 + \gamma \left(\frac{\omega L}{4\pi C_s}\right)^2\right\} \tag{11}$$

Application of the standing wave method to suppress flutes of $m > 1$ could be of interest for high-β stellarators, where $m = 1$ perturbations might be stable or only weakly unstable.

ACKNOWLEDGMENTS

The author should like to thank Drs. W. Grossmann and J. Junker for stimulating discussions and Dr. H. Weichselgartner for his help with the chemistry. He is further indebted to H. U. Scholz for assisting with the experiment and to G. Wendland and the Max-Planck-Institut für Extraterrestrische Physik for the use of the vibrator.

9.2 DYNAMIC STABILIZATION OF HELICAL AND SAUSAGE MODES IN ELECTRON-HOLE PLASMAS

B. Ancker-Johnson
Boeing Scientific Research Laboratories
Seattle, Washington 98124

ABSTRACT

Both helical (m > 0) and sausage (m = 0) instabilities, occurring spontaneously in injected electron-hole plasmas, are suppressed by rf energized magnetic quadrupoles of the Ioffe type. These modes oscillate at the same frequency, typically in the tens of MHz for plasmas in p-InSb, but possess different wavelengths. Suppression of >20 db is often achieved with rf currents (at 84 MHz in the present measurements) in the wall mirrors of less than 1A. The duration of the suppression is a function of the amplitude of rf current.

I. INTRODUCTION

Some effects of a magnetic trap of the type introduced by Gott, Ioffe, and Telkovskii (1963) on instabilities occurring in electron-hole plasmas were reported in 1964 (Ancker-Johnson, 1964). A schematic view down the axis of this device is shown in Fig. 1. The plasma is produced by electrical injection through contacts (not shown) covering the ends of the p-type InSb crystal. In addition to the four wall mirrors (or "Ioffe bars") indicated on the diagram by the plusses and minuses, there are two end mirrors composed of single turn coils (shaded) each of which receives half the current energizing the wall mirrors in the particular device discussed in this paper.

The previous report (Ancker-Johnson, 1964) showed that rf mirrors are more effective in stabilizing oscillations in electron-hole plasmas than are dc mirrors. Soon after rf mirrors were used also on fusion-type plasmas, again with agreeable results. (Orlinsky, Osovets, and Sinitsyn, 1965). An rf electric field has also been used to stabilize an electron-hole plasma (Dubovoi and Shanskii, 1965). The present paper reports more results obtained with the Ioffe quadrupole.

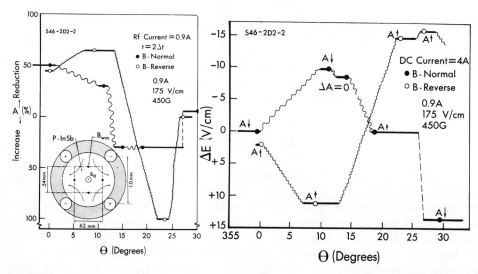

Fig. 1.   The dependence on angle between applied magnetic and electric fields of: (a) the amplitude of mixed mode oscillations during suppression by rf mirrors as observed at a time equal to twice the total suppression duration; and (b) the change in average electric field, for the same set of operating conditions, caused by dc mirrors. The normal and reverse directions of magnetic field refer to the axial component of the field only, not to the mirrors. The inset in (a) is an axial view of the quadrupole trap. The arrows in (b) indicate whether the amplitude of the oscillations increases or decreases when the dc mirrors are applied.

II.   THE UNSTABLE MODES

The helical instability ($m > 0$) has been extensively investigated in electron-hole plasmas (Ancker-Johnson, 1966). Recently the sausage mode ($m = 0$) has been identified and systematically studied (Ancker-Johnson, 1962; Drummond and Ancker-Johnson, 1964; Chen and Ancker-Johnson, 1970). The sausage mode is dominant in the absence of an axial magnetic field, although a few percent of the observed oscillation amplitude are caused by helical modes which, in turn, dominate in the presence of sufficiently strong axial magnetic fields. The dominant sausage mode occurs as a standing wave, whereas the $m = 1$ mode may occur as either a standing or a traveling wave. When more than one mode contributes to the observed oscillations, their frequencies are usually identical. The wavelength of the standing $m = 0$ mode is always twice the distance between anode and cathode; in the present device $\lambda = 8$mm. The $m = 1$ wavelength decreases with increasing plasma current (Chen and Ancker-Johnson, 1970).

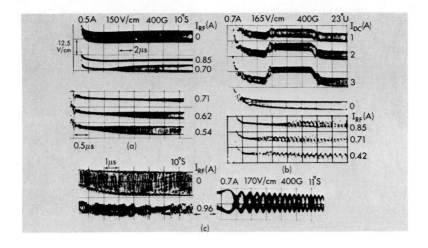

Fig. 2. Oscillogram showing the effects of mirrors on mixed mode oscillations. The vertical scale is the same for all traces and and the various horizontal scales are indicated. The operating conditions associated with (a), (b), and (c) are shown; they include plasma current, average electric field strength, magnitude of magnetic field and its direction. The notation S means the operating conditions produce oscillations which the dc mirror also stabilize, and U means the dc mirrors cannot stabilize.

The operating parameters which influence these instabilities are the plasma current and the applied magnetic field magnitude and direction. If, for example, the plasma current is held at 0.9 A (unpinched) and the magnitude of the magnetic field at 450G, Fig. 1a, and the angle between these is varied from parallel $(\theta = 0)$ to ~30°, several ranges in θ are found which correspond to regular oscillations (bars in Fig. 1). The θ-ranges in between (wavy lines in Fig. 1) correspond to noise-like oscillations. (Oscillations also occur for $\theta > 30°$ but they are beyond the scope of this paper.) These oscillations are usually the result of both sausage and helical instabilities. In the stabilization experiments described below no effort was made to isolate modes, thus m = 0 and m = 1 usually contributed comparable amplitudes to the observed oscillations.

III. RESULTS

Figure 2 shows some typical oscilloscope traces for operating conditions producing regular oscillations. The upper left trace

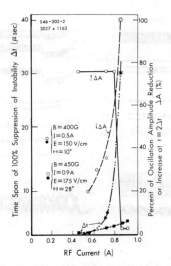

Fig. 3.   The rf mirror current influence on (1) the duration of total suppression (right-hand ordinant), and (2) the amplitude of the instability oscillations at a time equal to twice the duration of total suppression (left-hand ordinant).   The vertical arrows associated with the two amplitude curves indicate amplitude increase (open squares) or decrease (circles).   The point indicating 100% amplitude suppression is included to show that at 0.85A total suppression resulted for the longest time available in this measurement, 30μs.

results when no end or wall mirrors are present.   With just 0.85 A applied to the wall mirrors and half of that to the end mirrors all evidence of oscillations is gone.   When a somewhat smaller current is applied, 0.70A, the suppression lasts for 4μs.   The other oscillogram in Fig. 2a shows the progressively shorter duration of total suppression as the rf mirror current is reduced.   The dependence of this total suppression time on mirror current is shown in Fig. 3 (dots).   The amplitude of the oscillations at a time equal to twice the suppression time is, in this example, always less than the amplitude in the absence of mirrors, as the circles in Fig. 3 show.

The upper three oscillogram traces in Fig. 2b illustrate the influence of dc mirrors on the oscillations produced by another set of operating conditions.   These results show that the average voltage, and therefore, resistance of the plasma is increased as is the amplitude (slightly) of the oscillations.   So dc mirrors cause even more instability.   This is in contrast to the effect of dc mirrors on the first set of operating conditions, Fig. 2a (data not shown).   The lower oscillogram traces in Fig. 2b show that rf mirrors are much more effective in stabilization than dc, since the oscillations are completely suppressed even in weak mirrors (0.42A) for times long compared to a period of the instability oscillation, 29.4ns. The filled-in squares in Fig. 3 show the suppression time as a function of rf mirror current for oscillations which cannot be suppressed at all by dc mirrors.

At a time equal to twice the suppression time in the rf mirrors, the oscillations, which dc mirrors cannot stabilize at all, have a larger amplitude than the oscillations in the absence of mirrors (compare the four lowest traces in Fig. 2b). At this time the oscillations are also irregular in contrast to those which dc mirrors also suppress, Fig. la. An example of this amplitude dependence on rf mirror current is given by the open squares in Fig. 3.

A curious modulation is occasionally produced by the rf mirrors. A very large amplitude oscillation in the absence of mirrors is reproduced in Fig. 2c. The strongest mirrors possible with the equipment at hand, under a given set of operating conditions, suppress the oscillations by $\sim$8 db. A slight change in the operating conditions, in this case in θ from 10° to 11°, produces a modulation. The frequency of the modulation increases with time.

The oscillation amplitude is always suppressed by rf mirrors, for a time at least 2 orders of magnitude longer than the oscillation period. Beyond that time the amplitude may be more or less than the initial oscillations, as shown in Fig. la. In Fig. lb the effects of dc mirrors on oscillations produced by the same sets of operating conditions are shown for comparison. In some cases dc mirrors produce a lower electric field, a step toward stabilization, as indicated by the bars above the zero line. However, for only one of these sets of conditions does the dc mirrors also produce a decrease in oscillation amplitude, namely for θ = 10 to 12°. Thus, it is clear the rf mirrors are far more effective than dc mirrors in stabilizing the mixed modes of instability oscillations which occur spontaneously in electron-hole plasmas.

The two types of responses to rf mirrors summarized in Fig. 3 should be correlated with the instability mode nature. Are the easily stabilized oscillations (dots and circles in Fig. 3) dominantly sausage or helical modes, traveling or standing waves, with the converse being true of the harder-to-stabilize oscillation (open and closed squares)? Also, why is the rf mirror's effectiveness time dependent (Δt curves in Fig. 3)? Answers to such questions await more detailed studies.

9.3 DYNAMIC CONTROL OF CUSP LOSS

T. Sato, S. Miyake, T. Watari, T. Watanabe, S. Hiroe,
K. Takayama and K. Husimi
Institute of Plasma Physics, Nagoya University, Nagoya, Japan

ABSTRACT

Experimental results and computer simulation studies on the
dynamic control of cusp loss are presented. The steady-state
plasma is fed into a cusped magnetic field along the spindle axis.
The plasma density at the center is 10^{11} cm^{-3}, and $T_e \simeq T_i \simeq 10$ eV.
The rf electric field whose frequency is near the ion cyclotron frequency is applied perpendicularly to the line cusp and to both point
cusps. When the rf field is applied the ion density at the center
increases by a factor of two and its mean energy reaches up to 50 eV.
From the increase of ion density, the increase of decay time by
about a factor of ten is estimated. The number density of high-energy
ions above 100 eV confined in the cusp increases by a factor of over
fifty. The computer simulation of the rf effect on the line loss
suppression indicates a good agreement with the experimental
results.

I. INTRODUCTION

If there is some irreversible process that suppresses the particle loss and allows fuel plasma injection without any interference
at the feeding hole, it should appear promising for fusion reactors.
The dynamic control of a plasma at the feeding hole is proposed to
be a desired method of such a scheme. The containment of particles
with such a method has been investigated by applying an rf electric
field whose frequency is near the electron cyclotron frequency at
the leakage hole in open-ended systems (Consoli, 1969; Strijland,
1969; Eubank, 1969). Extensive investigation with ion cyclotron
frequency fields which has the advantage of direct ion heating, have
also been carried out[†](Watson and Kuo-Petravic, 1968; Miyake, Sato,
and Takayama, 1968). In the present work, the experimental results
are compared with those of computer simulation on the dynamic
control of cusp loss with the ion cyclotron resonance phenomenon.

[†] T. Sato, S. Miyake, T. Watari, Y. Kubota, T. Watanabe,
K. Takayama and K. Husimi, Annual Review of Institute of Plasma
Physics, Nagoya University, April,1969-March,1970, to be published.

II. EXPERIMENT

A steady state helium plasma with a diameter 4 mm is fed into
the cusp field along a spindle axis. The maximum field strength in
the line and the point cusp regions at 21 cm from center cusp, are
2.14 and 4.70 kG respectively. The neutral helium pressure in the
cusp chamber is 5×10^{-6} torr. Plasma density at the center ranges
from 10^{10} to 10^{11} cm^{-3}. An rf electric field is applied perpendicularly to the sheet plasma in the line cusp by two 6 cm apart parallel
electrodes whose outer and inner radii are 20 cm and 15 cm
respectively, as shown in Fig. 1. The other parallel plate
electrodes, 5 cm in length and 6 cm in width, are set up at both
point cusps.

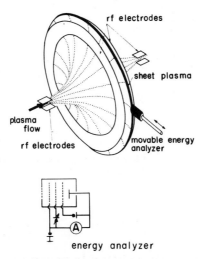

Fig. 1. Schematic of dynamic control system and retarding
potential type multigrid energy analyzer.

III. RESULTS

When the rf electric field is applied only at the line cusp, the
ion loss from the line cusp decreases at the resonance frequency
as shown in Fig. 2 and the ion density at the center of the cusp increases to almost twice that of no rf field. In the steady state
plasma, influx $F = (1/\tau_L + 2/\tau_P) N$, where τ_L and τ_P are the life
times governed by the line and point cusp losses respectively, and
N is the total number of ions in the containment region. Since the
line cusp loss is equal experimentally to twice the exit point cusp
loss, the suppression of the line cusp loss gives rise to a twofold

increase of N.   The result in Fig. 2 shows that $\tau'_L / \tau_L \gtrsim 10$, where τ'_L is the life time determined from the dynamically controlled line cusp loss. τ_L measured in a decaying plasma is about 400 μsec.[†] Consequently τ'_L is estimated at a few msec.

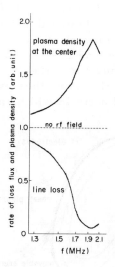

Fig. 2.   Rates of line loss and ion density at the center versus frequency. f_{ci} = 1.16 MHz, E_{rf} = 150 V/cm.

IV.   DISCUSSION

The present containment scheme is accompanied by effective ion heating due to the ion cyclotron damping of electrostatic wave. [†] The zero magnetic field point in the center makes the particle motion highly nonadiabatic, and consequently the energy of perpendicularly accelerated ions transfers to the parallel one.   The parallel energy of ions which leak from the line cusp is measured by the retarding potential type, multigrid energy analyzer movable along the sheet plasma.   Under the resonance condition, the contained ions consist of two components of energy both of which display Maxwellian distributions of $T_i \simeq$ 40 eV and 150 ~ 200 eV respectively when E_{rf} = 100 V/cm.   In Fig. 3 ion currents to the analyzer versus the radial distance for different retarding potential are shown.   When the retarding potential V_{ret} = -82 V, the ion flux increases in the central

---

[†] T. Sato, S. Miyake, T. Watari, Y. Kubota, T. Watanabe, K. Takayama and K. Husimi, Annual Review of Institute of Plasma Physics, Nagoya University, April, 1969-March, 1970, to be published.

region and decreases sharply at the outside of the rf electrodes to about 3% of that without rf field.

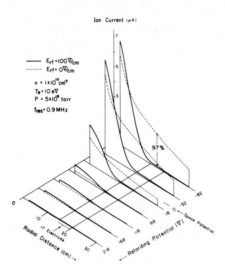

Fig. 3. Ion current to energy analyzer against radial distance from the center for different retarding potential.

The density of ions with energy higher than 100 eV increases in the entire region. The decrease of ion current at the location of the rf electrodes, however, shows the effective containment of these high energy ions. In addition, enlargement of the plasma volume is also observed. When rf fields are applied at the line and point cusps, the energy of contained ions reaches about 50 eV with a high energy component of above 200 eV.

Particle loss through the line cusp is studied by computer simulations in the same conditions as the experiment. The helium ions with a Maxwellian velocity distribution are fed into the cusp through a hole on the spindle axis. The rf electric field is uniform and directed along the axial direction having a cyclotron resonance zone at r = 18 cm. A particle is considered to be lost from the containment region, when $|z_i| >$ 20 cm or r > 22 cm. The energy distributions of the particles lost through the line cusp for different rf electric field are plotted in Fig. 4. The agreement between the experimental data and computer results indicates effective ion heating.

SATO et al.

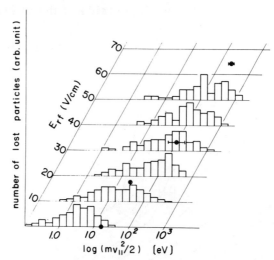

Fig. 4. Parallel energy distributions of ions lost through line cusp for different rf electric field. Results of computer simulations are shown with bar graphs and experimentally obtained ion temperature with ⊙.

The rf field strength in the plasma is estimated to be nearly equal to that in vacuum, for two reasons. First, the measured ion energy agrees with results of computer simulation based on the assumption that each ion is affected by the applied field strength; second, the thickness of the plasma sheet, across which the rf is applied, is about the size of an ion Larmor diameter. The density increase from ionization of the neutral gas (due to the applied rf field) is negligible. Even at resonance conditions, this density increase is estimated at about 10^5 cm^{-3}, since the ionization cross section is $\sim 10^{-18}$ cm^{-2}, the high-energy ion density $\sim 10^9$ cm^{-3}, the residual gas density $\sim 10^{11}$ cm^{-3}, and containment time ~ 100 μsec.

In conclusion, the application of dynamic control gives attractive results of twofold density increase in the central region and of effective ion heating. The energy of particles escaping from the line cusp is in good agreement with the result of computer simulations.

Acknowledgment

We wish to thank Dr. A. Outi of the Institute of Physical and Chemical Research for his extensive help of computation.

9.4 HOT ELECTRON CONFINEMENT IN A CUSPED MAGNETIC FIELD BY ECRH WRAPPING

S. Aihara, M. Fujiwara, M. Hosokawa and H. Ikegami

Institute of Plasma Physics, Nagoya University, Nagoya, Japan

ABSTRACT

The confinement of a hot electron plasma in a cusped magnetic field is studied. The hot electron plasma whose temperature is about 30 keV is produced in the electron cyclotron resonance region within the cusped magnetic field by a 20-msec pulse of 6.4 GHz having a power up to 5 kW. During the afterglow plasma, a weak 6.4 GHz microwave power is supplemented. Up to 150 W of supplemented power, the decay time increases linearly up to about 30 msec, about three times that without the supplement. With further increase of the power, the decay time decreases. The increase in the plasma containment time is explained to be due to the force $< F > = -(e^2/4m) \nabla \left[E^2/\omega^2 - \Omega_e^2 \right]$.

I. INTRODUCTION

Possibilities of confining thermonuclear plasma by radio frequency fields have been studied for hybrid combinations of static magnetic fields and rf fields, particularly for open-ended systems. If the applied radio frequency is near the cyclotron frequency, theory shows that the plasma loss is blocked by the stationary magnetic field gradient and the rf field in which the major load is borne by the magnetic field (Consoli, 1964; Canobbio, 1969). Trapping of electron beam in magnetic mirror geometry has been studied with some success (Consoli, 1969; Ikegami, Ikezi, Kawamura, Momota, Takayama, and Terashima, 1969; Dandle, Dunlap, Eason, Edmonds, England, Herrmann, and Lazar, 1969; Trivelpiece, Pechacek, and Kapetanakos, 1968; Eubank, 1969). In the present experiment, confinement is studied of a plasma generated and stably contained in a cusped magnetic field.

II. EXPERIMENT

The plasmas are produced in a cusped magnetic field by a 20-msec pulse of 6.4 GHz microwave, having a power up to 5 kW at a repetition rate of 10 pps. The microwave power is introduced into a vacuum chamber in such a way that the electric field is almost normal to the midplane and is in TM_{01}-like mode at the point cusp. The vacuum chamber is a stainless steel cylinder 17 cm long and 40 cm in diameter, being evacuated through a hole of 12 cm in

diameter on both end walls. Helium gas is continuously fed at
pressures ranging from 10^{-4} to 10^{-5} torr.

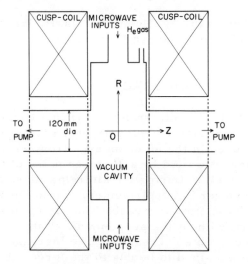

Fig. 1. Schematic diagram of the experimental device.

A schematic of the apparatus is shown in Fig. 1. The gradient
of the magnetic field intensity at the center is dB_z/dz (gauss/cm) is
equivalent to 410 I(kA)in the axial direction and dB_z/dv (gauss/
cm) = 172 I(kA)in the radial direction, where I is the cusp-coil
current. In order to locate a resonance contour of equi-magnetic-
field intensity within the chamber, at least I = 0.9 kA is required.

The plasma thus generated consists of hot electrons, cold
electrons, and cold helium ions. The hot-electron temperature is
determined to be about 30 keV from x-ray spectra and the density of
the hot electrons is about 10^{10} cm^{-3}, which is several percent of the
cold electron density. After removal of the microwave pulse, the
plasma density decreases to about 1/5 of the initial value within a
few msec and thereafter decays with a longer time constant of several tens of msec. The initial rapid decay is due to the cold electrons
and the successive slower decay to the hot electrons.

III. RESULTS

During the late afterglow plasma, a weak microwave power is
supplemented at the same frequency as the main discharging power.
For up to 150 W of the supplemented power, the decay time, determined from a semi-log plot of the hot-electron density versus time,

is observed to increase linearly up to a value about three times that without supplement. With further increase of the power, the decay time starts decreasing. Typical experimental results are shown in Fig. 2.

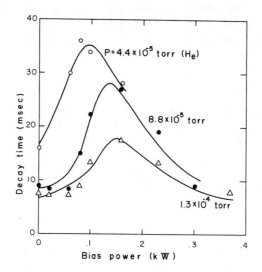

Fig. 2. Decay time of the hot electrons as a function of bias-power after the turnoff of main discharging power 2 kW. Cusped magnetic field intensity is dB_r/dr = 172 I (kA) gauss/cm and dB_z/dz = 410 I (kA) gauss/cm.

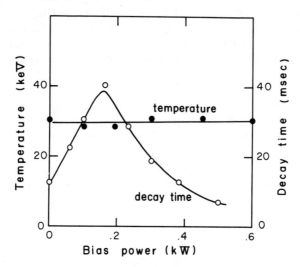

Fig. 3. Temperature and decay time of the hot electrons as a function of bias-power after the turnoff of main discharging power 3 kW.

It is observed that the leveling off of the decay time appears at
smaller bias-power for lower gas pressure. The bias power is so
small that heating of electrons is negligible as shown in Fig. 3.
Decreasing of the decay time with the supplemented power beyond
a certain critical value is attributed to certain discharge effects,
which may also limit the maximum attainable density in high-power,
ECRH plasma experiments. Since no appreciable fluctuations are
detected, the decrease in decay time cannot be understood as being
caused by enhanced diffusion loss.

IV. DISCUSSION

The increase in the plasma containment time is explained to be
due to the local, electron cyclotron resonance effect, that is, the
electrons escaping from the cusp are pushed back by the force

$$< F > \; = \; -\frac{e^2}{4m} \; \nabla \; \frac{E^2}{\omega^2 - \Omega_e^2} \; . \tag{1}$$

Since the resonance region in the cusped geometry forms a closed
surface, the containment effect may be called 'ECRH wrapping'.
Although the local electric field is not so large as to drive back the
30 keV electrons directly, the electric field is still large enough
to correct small-angle deflections. During the stable decay most of
the escaping electrons are gradually generated by multiple small-
angle scatterings. From the typical decay time of 10 msec without
supplement microwave field, it is quite likely that the electrons are
trapped in a mirror-like region of the cusped magnetic field. By
using a Langmuir probe, the density distribution is measured; as
shown in Fig. 4 the plasma is observed to form a shell-structure
and the density is minimum at the center where the magnetic field
intensity is zero.

In conclusion, it is confirmed that the plasma can be confined
in a cusped magnetic field by ECRH wrapping. It should be mentioned, however, that the effect of ECRH wrapping is observed to
be more remarkable in the cusped magnetic field than in the mirror
field. The decrease of the decay time with supplemented power
beyond a certain critical value is not yet well understood, this effect
will limit rf confinement.

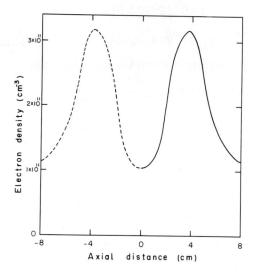

Fig. 4. Axial density distribution of cold-electron component which forms a shell-structure centered on the axis of the machine, with helium pressure 4×10^{-5} torr and discharging microwave power at 2 kW. Radial profile of cold electrons is observed also to have a shell-structure.

Acknowledgments

The authors are grateful to Professor K. Takayama for his interest and encouragement, who suggested the study of ECRH wrapping.

BIBLIOGRAPHY

Abrams, R. H. Jr., E. J. Yadlowsky, and H. Lashinsky (1969). Periodic pulling and turbulence in a bounded plasma. Phys. Rev. Letters $\underline{22}$, 275.

Alcock, M. W. and B. E. Keen (1970). Experimental observation of the drift-dissipative instability in afterglow plasmas. J. Phys. A $\underline{3}$, L21.

Ancker-Johnson, B. (1962). Observations of growing oscillations in electron-hole plasma. Phys. Rev. Letters $\underline{9}$, 485.

Ancker-Johnson, B. (1964). Stabilization of electron-hole plasmas by a magnetic trap. Phys. Fluids $\underline{7}$, 1553.

Ancker-Johnson, B. (1966). Plasmas in semiconductors and semimetals. Semiconductors and Semimetals (Academic Press, New York) Vol. 1, p. 379.

Ancker-Johnson, B. (1968). Plasma effects in semiconductors. Proc. Intern. Conf. on Physics of Semiconductors (Nauka, Leningrad) p. 813.

Ancker-Johnson, B. (1970). Dynamic stabilization of helical and sausage modes in electron-hole plasmas. These Proceedings.

Ancker-Johnson, B., H. J. Fossum, and A. Y. Wong (1970). Feedback stabilization and measurement of instability coefficients in electron-hole plasmas. These Proceedings.

Anderson, O. A., D. H. Birdsall, C. W. Hartman, E. J. Lauer, and H. P. Furth (1969).
Plasma confinement in the levitron. Plasma Physics and Controlled Nuclear Fusion Research (International Atomic Energy Agency, Vienna) Vol. 1, p. 443.

Arsenin, V. V. (1969).
The possibility of suppression of helical modes of hydromagnetic instability of a plasma column by a feedback system. International Symposium on Closed Confinement Systems, Dubna. (To be published).

Arsenin, V. V. (1970).
Stabilization of large-scale plasma instabilities by feedback. Zh. Tekh. Fiz. 39, 1553. Sov. Phys. --Tech. Phys. 14, 1166.

Arsenin, V. V. and V. A. Chuyanov (1968).
A possible stabilization of the trough-shaped plasma instability through a feedback system. Dokl. Akad. Nauk 180, 1078. Sov. Phys.--Dokl. 13, 570.

Arsenin, V. V., V. A. Zhiltsov, V. Kh. Likhtenstein, and V. A. Chuyanov (1968).
Suppression of cyclotron instability of a rarified plasma with the aid of a feedback system. ZhETF Pis. Red. 8, 69. JETP Lett. 8, 41.

Arsenin, V. V., V. A. Zhiltsov, and V. A. Chuyanov (1969).
Suppression of flute instability of a plasma by means of a feedback system. Plasma Physics and Controlled Nuclear Fusion Research (International Atomic Energy Agency, Vienna) Vol. 2, p. 515.

322

Artemenkov, L. I., R. P. Vasilev, I. V. Galkin, I. H. Golovin,
V. A. Zhilt́sov, V. F. Zubarev, O. N. Kachalov, N. I.
Klochkov, V. P. Konyaev, V. V. Kuznetsov, V. Kh.
Likhtenstein, E. A. Maslennikov, P. A. Mukhin,
A. N. Nekrasov, D. A. Panov, V. I. Pistunovich,
Yu. M. Pustovoit, V. S. Svishchev, N. N. Semashko,
A. A. Tereshkin, K. Z. Tushabramishvili, V. I.
Chuyanov, I. A. Chukhin, and V. I. Yushkevich (1966).
Production of a hot plasma in OGRA-II by the injection
of fast hydrogen atoms into a magnetic trap. Plasma
Physics and Controlled Nuclear Fusion (International
Atomic Energy Agency, Vienna) Vol. 2, p. 45.

Artsimovitch, L. A. (1964).
Controlled Thermonuclear Reactions (Gordon and
Breach Publishers, New York) p. 351.

Artsimovitch, L. A., G. A. Bobrovskii, E. P. Gorbunov,
D. P. Ivanov, B. D. Kirilov, E. I. Kuznetsov,
S. V. Mirnov, M. P. Petrov, K. A. Razumova,
V. S. Strelkov, and D. A. Shcheglov (1969).
Experiments in Tokamak Devices. Plasma Physics
and Controlled Nuclear Fusion Research (International
Atomic Energy Agency, Vienna) Vol. 1, p. 157.

Berezin, A. K., Ya. B. Fainberg, L. I. Bolotin, G. P. Berezina,
I. A. Bezyazychny, V. I. Kirilko, V. D. Shapiro, and
V. P. Zeidlits.(1969).
Control of beam instabilities. Plasma Physics and
Controlled Nuclear Fusion Research (International
Atomic Energy Agency, Vienna) Vol. 2, p. 723.

Berge, G. (1968).

    Dynamic stabilization of unstable MHD equilibria,
a general approach. Culham Laboratory Report
CLM-R97.

Berge, G. (1969).

    On the problem of dynamic stabilization. <u>Plasma Physics
and Controlled Nuclear Fusion Research</u> (International
Atomic Energy Agency, Vienna) Vol. 2, p. 483.

Berge, G. (1970).

    Dynamic stabilization of the m=1 instability in a bumpy
theta pinch. Phys. Fluids <u>13</u>, 1031.

Bernstein, I. B., E. A. Frieman, M. K. Kruskal, and
R. M. Kulsrud (1958).

    An energy principle for hydromagnetic stability
problems. Proc. Phys. Soc. (London) <u>244A,</u> 17.

Blank, A. A., H. Grad, and H. Weitzner (1969).

    Toroidal high-β equilibria. <u>Plasma Physics and
Controlled Nuclear Fusion Research</u> (International
Atomic Energy Agency, Vienna) Vol. 2, p. 607.

Bobrovskii, G. A., K. A. Razumova, and D. A. Shcheglov (1968),
Some pecularities of the plasma behavior in Tokamak
TM-3. Plasma Physics <u>10</u>, 436.

Bobyrev, N. A. and O. I. Fedyanin (1963).

    High frequency magnetic field stabilization of a cylinder
carrying current. Zh. Tekh. Fiz. <u>33</u>, 1187. Sov.
Phys.--Tech. Phys. <u>8</u>, 887.

Bodin, H. A. B., E. P. Butt, J. McCartan, and G. H. Wolf (1969).
Dynamic stabilization of an m=1 instability. Third
European Conference on Controlled Fusion and Plasma
Physics (Wolters-Noordhoff Publishing, Groningen
Netherlands) p. 76.

Bodin, H. A. B. and J. McCartan (1969).
High beta toroidal experiments. International Symposium
on Closed Confinement Systems Dubna. To be published.

Bodin, H. A. B, J. McCartan, A. A. Newton, and G. H. Wolf (1969).
Diffusion and stability of high-β plasma in an 8-metre
theta pinch. Plasma Physics and Controlled Nuclear
Fusion Research (International Atomic Energy Agency,
Vienna) Vol. 2, p. 533.

Braginskii, S. I. (1965).
Transport processes in a plasma. Reviews of Plasma
Physics M. A. Leontovich, Ed. (Consultants Bureau
Enterprises, Inc., New York) Vol. 1, p. 205.

Briggs, R. J. (1964).
Criteria for identifying amplifying waves and absolute
instabilities. Electron Stream Interaction with Plasmas
(M. I. T. Press, Cambridge) p. 8.

Brown, S. C. (1959).
Basic Data of Plasma Physics (John Wiley and Sons,
Inc., New York).

Buchelnikova, N. S. (1964).
Universal instability in a potassium plasma. Zh. Eksp.
Teor. Fiz. 46, 1147. Sov. Phys.--JETP 19, 775.

Canales, R. (1965).

> Stability of the hydromagnetic pinch with ideal
> feedback. Master's Thesis, Dept. of Electrical
> Engineering, M.I.T., Cambridge.

Canobbio, E. (1969).

> Gyroresonant particle acceleration in a non-uniform
> magnetostatic field. Nucl. Fusion 9, 27.

Carlyle, D. C. (1970).

> Passive feedback stabilization. These Proceedings.

Chandrasekhar, S. (1961).

> Hydrodynamic and Hydromagnetic Stability (Clarendon
> Press, Oxford, England).

Chang, R. P. H. and M. Porkolab (1970).

> Back-scattering instability of cyclotron harmonic
> waves in nonlinear decay processes. Princeton
> University Plasma Physics Laboratory Report
> MATT-750.

Chen, F. F. (1966).

> Microinstability and shear stabilization of a low-β,
> rotating, resistive plasma. Phys. Fluids 9, 965.

Chen, F. F. and C. Etievant (1970).

> Convection arising from large-amplitude plasma waves.
> Phys. Fluids 13, 687.

Chen, F. F. and H. P. Furth (1969).

> Low frequency plasma stabilization by feedback
> controlled neutral beams. Nucl. Fusion 9, 364.

Chen, W. S. and B. Ancker-Johnson (1970).
Pinch oscillations in electron-hole plasmas.
Boeing Scientific Research Laboratories Report
D1-82-0965.

Chodura, R. (1964).
Zum Zündmechanismus einer Theta-Pinch-Entladung.
Z. Naturforsch. 19A, 679.

Chu, T. K., B. Coppi, H. W. Hendel, and F. W. Perkins (1969).
Drift instabilities in a uniformly rotating plasma
cylinder. Phys. Fluids 12, 203.

Chu, T. K., H. W. Hendel, D. L. Jassby, and T. C. Simonen
(1970). Feedback stabilization of the transverse Kelvin-
Helmholtz instability--experiment and theory. These
Proceedings.

Chu, T. K., H. W. Hendel, R. W. Motley, F. W. Perkins,
P. A. Politzer, T. H. Stix, and S. von Goeler (1969).
Finite ion Larmor radius and ion-ion collision effects
on equilibrium, critical fluctuation and drift wave
states of alkali-metal plasmas. Plasma Physics and
Controlled Nuclear Fusion Research (International
Atomic Energy Agency, Vienna) Vol. 1, p. 611.

Chu, T. K., H. W. Hendel, and P. A. Politzer (1967).
Measurements of enhanced plasma losses caused by
collisional drift waves. Phys. Rev. Letters 14, 1110.

Chu, T. K., H. W. Hendel, L. G. Schlitt, T. C. Simonen,
and T. H. Stix (1969). Soft-and hard-onset and
amplitude saturation of collisional drift instabilities.
Proceedings of Conference on Physics of Quiescent
Plasmas (Ecole Polytechnique, Paris) Pt. II, p. 57.

Chu, T. K., H. W. Hendel, and T. C. Simonen (1970).
Spontaneous decay mode of drift instability.
Bull. Am. Phys. Soc. 15, 777.

Chuyanov, V. A. (1969).
Experimental results on the suppression of the flute
instability by a feedback system on Phoenix II.
Culham Laboratory Report CTO/598.

Chuyanov, V. A., E. G. Murphy, D. R. Sweetman, and
E. Thompson (1969). Suppression of the flute
instability by a feedback system on Phoenix II.
Bull. Am. Phys. Soc. 14, 1028.

Consoli, T. (1964).
Recentes études sur les décharges H. F. non controlées par le
regime de diffusion. Proceedings of the Sixth International
Conference on Ionization Phenomena in Gases P. Hubert and
E. Cremieu-Alcan Eds. (Bureau des Editions, Centre
d'Etudes Nucleaires de Saclay, Paris) Vol. 2, p. 455.

Consoli, T. (1969).
Theoretical and experimental studies of high frequency
plasma structure. Plasma Physics and Controlled
Nuclear Fusion Research (International Atomic Energy
Agency, Vienna) Vol. 2, p. 361.

Cotsaftis, M. (1968).
Dynamical stabilization of low-frequency microinstabilities. Phys. Letters 27A, 662.

328

Crowley, J. M. (1967).

Stabilization of a spatially growing wave by feedback.
Phys. Fluids 10, 1170.

Dandl, R. A., J. L. Dunlap, H. O. Eason, P. H. Edmonds,
A. C. England, W. J. Herrmann, and N. H. Lazar (1969).
Electron cyclotron heated "target" plasma experiments.
Plasma Physics and Controlled Nuclear Fusion Research
(International Atomic Energy Agency, Vienna)
Vol. 2, p. 435.

Davydov, A. S. (1965).

Quantum Mechanics (Addison-Wesley Publishing Co.,
Reading, Mass.).

Dawson, J. and C. Oberman (1962).

The high frequency conductivity and the emission and
absorption coefficients of a fully ionized plasma.
Phys. Fluids 5, 517.

Dawson, J. and C. Oberman (1963).

Effect of ion correlations on the high frequency
conductivity and emmisssion and absorption coefficients
of a fully ionized plasma. Phys. Fluids 6, 394.

Demirkhanov, R. A. (1968).

Stabilisation d'un plasma par des champs électro-
magnetiques de haute fréquence. Deuxieme Colloque
International sur les Interactions Champs Oscillants
Plasmas (Commisariat d'Energie Atomique, Direction
de la Physique, Fontenay-aux-Rose, France) Vol. III,
p. 809.

Dewan, E. M. and H. Lashinsky (1969).
Asynchronous quenching of the van der Pol oscillator.
IEEE Trans. Automatic Control 14, 212.

Dobrowolny, M., F. Engelmann, and A. M. Levine (1969a).
Dynamic stabilization of drift waves by electron
temperature modulation. Plasma Physics 11, 973.

Dobrowolny, M., F. Engelmann, and A. M. Levine (1969b).
Dynamic stabilization of collision-dominated drift
waves in the presence of an a.c. current. Plasma
Physics 11, 983.

Dressler, J. L. (1970).
Video type sampling in the feedback stabilization
of electromechanical equilibria. These Proceedings.

Drummond, J. E., R. A. Gerwin, and B. G. Springer (1961).
Concept of conductivity. J. Nucl. Energy Pt. C,
2, 98.

Drummond, J. E. and B. Ancker-Johnson (1964).
Theory of pinch effect in electron-hole plasmas.
Proceedings of the Seventh International Conference
on Physics of Semiconductors (Academic Press,
New York) Vol. 2, p. 173.

Dubovoi, L. V. and V. F. Shanskii (1965).
Stabilization of a helical instability in an electron-
hole plasma in a semiconductor by an alternating
electric field. Zh. Eksp. Teor. Fiz. 48, 800.
Sov. Phys.--JETP 21, 530.

Dupree, T. H. (1967).

> Nonlinear theory of drift-wave turbulence and enhanced diffusion. Phys. Fluids $\underline{10}$, 1049.

Ellis, R. A. and M. Porkolab (1968).

> Nonlinear interactions of cyclotron harmonic waves. Phys. Rev. Letters $\underline{21}$, 529.

Etievant, C., S. Ossakow, E. Ozizmir, C. H. Su, and I. Fidone (1968). Nonlinear interaction of electromagnetic waves in a cold magnetized plasma. Phys. Fluids $\underline{11}$, 1778.

Eubank, H. P. (1969).

> Single-particle confinement studies in a radio-frequency-supplemented magnetic mirror. Phys. Fluids $\underline{12}$, 234.

Fainberg, Ya. B. and V. D. Shapiro (1967).

> Drift instabilities of a plasma situated in a high-frequency electric field. Zh. Eksp. Teor. Fiz. $\underline{52}$, 293. Sov. Phys.--JETP $\underline{25}$, 189.

Fedorchenko, V. D., V. I. Muratov, and B. N. Rutkevich (1964). High-frequency oscillations of a plasma in a magnetic field. Zh. Tekh. Fiz. $\underline{34}$, 458. Sov. Phys.-- Tech. Phys. $\underline{9}$, 358.

Freidberg, J. P. (1969).

> Magnetohydrodynamic stability of a diffuse screw pinch. Los Alamos Scientific Laboratory Report LA-DC-10793.

Fuenfer, E., J. Junker, M. Kaufmann, J. Neuhauser, and
V. Seidel (1968). Influence of a bump on the stability
of a high-energy theta-pinch plasma. Bull. Am. Phys.
Soc. 13, 1552.

Furth, H. P., M. N. Rosenbluth, P. H. Rutherford, and
W. Stodiek (1970). Thermal equilibrium and stability
of Tokamak discharges. Princeton University
Plasma Physics Laboratory Report MATT-778.

Furth, H. P. and P. H. Rutherford (1969).
Feedback stabilization of drift waves by modulated
electron sources. Phys. Fluids 12, 2638.

Garscadden, A. and P. Bletzinger (1969).
Feedback stabilization and mode-coupling of
ionization waves. Conference on Physics of Quiescent
Plasmas (Ecole Polytechnique, Paris) Pt. II, p. 99.

Gelb, A. and W. E. Van der Velde (1968).
Multiple-Input Describing Functions and Nonlinear
System Design (McGraw-Hill Book Co., New York).

Gott, I. V., M. S. Ioffe, and V. G. Telkovskii (1962).
New results concerning plasma in magnetic traps.
Nucl. Fusion Supp. Pt. 3, 1045.

Gould, L. A. (1969).
Chemical Process Control: Theory and Applications
(Addison-Wesley Publishing Co., Reading, Mass.)
p. 235.

Grad, H. and H. Weitzner (1969).
Critical β stellarator and scyllac expansions. Phys.
Fluids 12, 1725.

Gruber, O., H. Herold, R. Wilhelm, and H. Zwicker (1969).
MHD instabilities in a toroidal and a linear high-β
screw pinch. Bull. Am. Phys. Soc. 14, 1031.

Haas, F. A. and J. A. Wesson (1966).
Stability of the theta-pinch. Phys. Fluids 9, 2472.

Haas, F. A. and J. A. Wesson (1967a).
Dynamic stabilization of the theta-pinch. Phys. Rev.
Letters 19, 833.

Haas, F. A. and J. A. Wesson (1967b).
Stability of the theta-pinch II. Phys. Fluids 10, 2245.

Hartman, C. W., R. H. Munger, and M. F. Uman (1969).
Plasma confinement in hybrid-multipole-stellarator
fields. Lawrence Radiation Laboratory Report
UCRL-71721.

Hasegawa, A. (1968).
Theory of longitudinal plasma instabilities. Phys.
Rev. 169, 204.

Hedrick, C. L. Jr. (1970).
Interaction of quasitransverse and quasilongitudinal
waves in an inhomogeneous plasma. Ph. D. Thesis,
University of California.

Hendel, H. W., T. K. Chu, and P. A. Politzer (1968).
Collisional drift waves--identification, stabilization,
and enhanced plasma transport. Phys. Fluids 11, 2426.

Hendel, H. W., T. K. Chu, F. W. Perkins, and T. C. Simonen
(1970). Remote feedback stabilization of collisional
drift instability by modulated microwave energy
source. Phys. Rev. Letters 24, 90.

Hockney, R. W. (1968).
    Characteristics of noise in a two-dimensional
    computor plasma. Phys. Fluids 11, 1381.

Ikegami. H., H. Ikezi, T. Kawamura, H. Momota, K. Takayama,
    and Y. Terashima (1969). Characteristics of micro-
    instabilities in a hot-electron plasma. Plasma Physics
    and Controlled Nuclear Fusion Research (International
    Atomic Energy Agency, Vienna) Vol. II, p. 423.

Ivanov, A. A., Yu. B. Kazakov, A. N. Lukyanchuk, V. D Rusanov,
    and S. S. Sobolev (1969). Stabilization of potential
    oscillations of a plasma in a Q machine by means of
    a high frequency magnetic field. ZhETF Pis. Red.
    9, 356. JETP Lett. 9, 210.

Ivanov, A. A., L. I. Rudakov, and J. Teichmann (1968).
    Effect of a high-frequency magnetic field on plasma
    instability. Zh. Eksp. Teor. Fiz. 54, 1380.
    Sov. Phys.--JETP 27, 739.

Ivanov, A. A. and J. Teichmann (1969).
    Stabilization of some dissipative instabilities by means
    of high-frequency electromagnetic field. Czech. J.
    Phys. 19B, 941.

Jassby, D. L. and F. W. Perkins (1970).
    Transverse Kelvin-Helmholtz instability in a Q machine
    plasma. Phys. Rev. Letters 24, 256.

Jury, E. I. (1964).
    Theory and Application of the Z Transform Method
    (John Wiley and Sons, Inc., New York) p. 2.

Kadomtsev, B. B. (1965).

Plasma Turbulence (Academic Press, New York).

Kawabe, T. (1966).

Electron plasma oscillations excited by two-beam
instability and nonlinear coupling between them.
J. Phys. Soc. Japan 21, 2704.

Keen, B. E. (1970).

Interpretation of experiments on feedback control of
a "drift-type" instability. Phys. Rev. Letters 24, 259.

Keen, B. E. and R. V. Aldridge (1969).

Suppression of a "drift-type" instability in a magneto-
plasma by a feedback technique. Phys. Rev. Letters
22, 1358.

Keen, B. E. and W. H. W. Fletcher (1969).

Suppression and enhancement of an ion-sound instability
by nonlinear resonance effects in a plasma. Phys.
Rev. Letters 23, 760.

Keen, B. E. and W. H. W. Fletcher (1970).

Suppression of a plasma instability by the methods of
"asynchronous quenching." Phys. Rev. Letters
24, 130.

Kent, G. I., N. C. Jen, and F. F. Chen (1969).

Transverse Kelvin-Helmholtz instability in a rotating
plasma. Phys. Fluids 12, 2140.

Kornilov, E. A., Ya. B. Fainberg, L. I. Bolotin, and O. F.
Kovpik (1966). Suppression of low-frequency oscillations in two-stream instabilities by prior modulation
of the electron beam. ZhETF Pis. Red. 3, 354.
JETP Lett. 3, 229.

Kotowski, G. (1943).

Loesungen der inhomogenen Mathieuschen Differential-
gleichung mit periodischer Stoerfunktion beliebiger
Frequenz. Z. Angew. Math. Mech. 23, 213.

Kubo, R. (1966).

The fluctuation-dissipation theorem. Rept. Progr.
Phys. 29, 255.

Kuckes, A. F. and J. M. Dawson (1965).

Electron cyclotron harmonic radiation from a plasma.
Phys. Fluids 8, 1007.

Kuehl, H. (1967).

Coupling of transverse and longitudinal waves below
the second electron cyclotron harmonic. Phys. Rev.
154, 124.

Landau, L. D. and E. M. Lifshitz (1959).

Fluid Mechanics (Addison-Wesley Publishing Co.,
Reading, Mass.). p. 104.

Landau, L. D. and E. M. Lifshitz (1960a).

Electrodynamics of Continuous Media (Addison-
Wesley Publishing Co., Reading, Mass.). p. 256.

Landau, L. D. and E. M. Lifshitz (1960b).

Mechanics (Addison-Wesley Publishing Co., Reading
Mass.).

Lashinsky, H (1965).

Nonlinear mode interactions in universal plasma
instabilities. Phys. Rev. Letters 14, 1065.

336

Lashinsky, H. (1966).

Universal instability in a thermal plasma device
(Q machine). Plasma Physics and Controlled Nuclear
Fusion Research (International Atomic Energy Agency,
Vienna) Vol. 1, p. 499.

Lashinsky, H. (1969).

Mathematical models for nonlinear mode interaction.
Nonlinear Effects in Plasmas (Gordon and Breach,
New York). G. Kalman and M.R. Feix, edts.

Lashmore-Davies, C. N. (1970).

Stabilization of a low density plasma in a simple
magnetic mirror by feedback control. These Proceedings

Lee, D. A. , P. Bletzinger, and A. Garscadden (1966).
Wave nature of moving striations. J. Appl. Phys.
37, 377.

Lepechinsky, D. and P. Rolland (1969).
Stabilisation par un champ de haute fréquence de
l'instabilité de dérive dissipative. Compt. Rend.
268B, 957.

Lighthill, M. J. (1965).
Contributions to the theory of waves in nonlinear
dispersive systems. J. Inst. Math. Appls. 1, 269.

Lindgren, N. E. and C. K. Birdsall (1970).
Feedback suppression of collisionless, multimode
drift waves in a mirror-confined plasma. Phys.
Rev. Letters 24, 1159.

Lindman, E. L. (1967).

    Effect of generalized boundary conditions on drift
waves. Bull. Am. Phys. Soc. 12, 1136.

Lisitano, G. (1966).

    Production of a high-frequency plasma using a slotted
cylinder system. Proceedings of the Seventh
International Conference on Phenomena in Ionized
Gases (Gradevinska Knjiga Publishing House, Beograd)
Vol. 1, p. 464.

Little, E. M., A. A. Newton, W. E. Quinn, and F. L. Ribe
(1969). Linear theta-pinch experiments related to the
stability of a toroidal theta-pinch of large aspect ratio.
Plasma Physics and Controlled Nuclear Fusion
Research (International Atomic Energy Agency, Vienna)
Vol. II, p. 555.

Little, P. F. and P. J. Barrett (1967).

    Drift waves in a cesium plasma. Conference on
Physics of Quiescent Plasmas (Laboratori Gas
Ionizzati, Frascati, Italy) Pt. I, p. 173.

Lord Rayleigh (1900).

    Investigation of the character of the equilibrium of an
incompressible heavy fluid of variable density.
Scientific Papers (Cambridge University Press,
Cambridge, England) Vol. II, p. 200.

Lotz, W., E. Remy, and G. H. Wolf (1964).

    Toroidal theta-pinch in M-and-S configuration.
Nucl. Fusion 4, 335.

Margenau, H. (1946).

>   Conduction and dispersion of ionized gases at high
>   frequencies. Phys. Rev. 69, 508.

Meixner, J. and F. W. Schaefke (1954).

>   Mathieusche Funktionen und Sphaeroidfunktionen.
>   (Springer-Verlag, Berlin).

Melcher, J. R. (1965a).

>   Control of a continuum electromechanical instability.
>   Proc. IEEE 53, 460.

Melcher J. R. (1965b).

>   An experiment to stabilize an electromechanical con-
>   tinuum. IEEE Trans. Auto. Contr. AC-10, 466.

Melcher, J. R. (1966).

>   Continuum feedback control of instabilities on an
>   infinite fluid interface. Phys. Fluids 9, 1973.

Melcher, J. R. (1968).

>   Complex waves. IEEE Spectrum 5, 86.

Melcher, J. R. (1969).

>   M. I. T., Center for Space Research Progress Report
>   CSR-TR-69-12.

Melcher, J. R. (1970).

>   Feedback stabilization of hydromagnetic continua:
>   review and prospects. These Proceedings.

Melcher, J. R. and E. P. Warren (1966).

>   Continuum feedback control of a Rayleigh-Taylor
>   type instability. Phys. Fluids 9, 2085.

Meyer, F. and H. U. Schmidt (1958).

Torusartige Plasmakonfigurationen ohne Gesamtstrom durch ihren Querschnitt in Gleichgewicht mit einem Magnetfeld. Z. Naturf. 13A, 1005.

Mikhailovsky, A. B. (1966).

Nonlinear theory of drift-cone instability. Dokl. Akad. Nauk 169, 554. Sov. Phys.--Dokl. 11, 603.

Millner, A. R. and R. R. Parker (1970).

Nonlinear stabilization of a continuum. These Proceedings.

Minorsky, N. (1962).

Nonlinear Oscillations (D. Van Nostrand Co. Inc., Princeton).

Mirnov, S. V. and I. B. Semenov (1969).

Correlational method of studying plasma instabilities in the Tokamak-3 apparatus. Kurchatov Institute of Atomic Energy Report IAE-1907. Trans. Oak Ridge National Laboratory Report ORNL-tr-2352.

Miyake, S., T. Sato, and K. Takayama (1967).

Acceleration of ions in a plasma by using an RF electric field in a non-uniform magnetic field. Annual Review of the Institute of Plasma Physics, Nagoya, Japan p. 63.

Mori, H. (1965).

Transport collective motion and brownian motion. Progr. Theor. Phys. 33, 423.

Morozov, A. I. and L. S. Solovev (1964).

Cybernetic stabilization of plasma instabilities. Zh. Tekh. Fiz. 34, 1566. Sov. Phys.--Tech. Phys. 9, 1214.

Morse, R. L., W. B. Risenfeld, and J. L. Johnson (1968).
Hydromagnetic equilibrium of a thin skin finite
beta toroidal plasma column. Plasma Physics
10, 543.

Mosher, D. and F. F. Chen (1970).
Convective losses in a thermionic plasma with shear.
Phys. Fluids 13, 1328.

Mueller, G., W. Friz, and R. S. Palmer (1970).
Measurements of growth and damping rates in an
inhomogeneous low-β plasma column. Aerospace
Research Laboratories, Report No. ARL 70-0080.

Mueller, G., J. C. Corbin, and R. S. Palmer (1970a).
Characteristics of a modified feedback method applied
to edge oscillation in a Q machine. These Proceedings.

Mueller, G., J. C. Corbin, and R. S. Palmer (1970b).
Characteristics of low frequency oscillations near
the edge of a low-β plasma column. Bull. Am.
Phys. Soc. 15, 815.

Nishida, Y., M. Tanibayashi, and K. Ishii (1970).
Suppression of drift-wave instability by RF electric
field. Phys. Rev. Letters 24, 1001.

Orlinski, D. V., S. M. Osovets, and V. I. Sinitsyn (1965).
Dynamic stabilization of a plasma column. Plasma
Physics and Controlled Nuclear Fusion Research
(International Atomic Energy Agency, Vienna)
Vol. II, p. 313.

Osovets, S. M. (1960).

    Dynamic stabilization of a plasma ring. Zh. Eksp.
Teor. Fiz. <u>39</u>, 311. Sov. Phys.--JETP <u>12</u>, 221.

Osovets, S. M. and V. I. Sinitsyn (1965).

    Dynamic stabilization of a plasma pinch. Zh. Eksp.
Teor. Fiz. <u>48</u>, 1171. Sov. Phys.--JETP <u>21</u>, 715.

Parker, R. R. and K. I. Thomassen (1969).

    Feedback stabilization of a drift-type instability.
Phys. Rev. Letters <u>22</u>, 1171.

Pavlov, E. I. and V. I. Sinitsyn (1966).

    Suppression of instabilities in a pinch. Zh. Eksp.
Teor. Fiz. <u>51</u>, 87. Sov. Phys. JETP <u>24</u>, 59.

Pekarek, L. (1968).

    Ionization waves (striations) in a discharge plasma.
Usp. Fiz. Nauk <u>94</u>, 463. Sov. Phys.--Usp. <u>11</u>, 188.

Pengilley, C. J. and P. M. Milner (1967).

    On asynchronous quenching. IEEE Trans. Prof. Group.
Auto. Control <u>12</u>, 224.

Perkins, F. W. and D. L. Jassby (1970).

    Velocity shear and low-frequency plasma instabilities.
Princeton University Plasma Physics Laboratory
Report MATT-751

Pfirsch, D. and H. Wobig (1966).

    Equilibrium and stability of the M-and-S torus and
related configurations. <u>Plasma Physics and Controlled
Nuclear Fusion Research</u> (International Atomic Energy
Agency, Vienna) Vol. 1, p. 757.

Politzer, P. A. (1969).

Drift instability in collisionless alkali metal plasmas.
Ph. D. Thesis, Princeton University.

Rébut, P. H. and A. Samain (1969a).

Variable angulaires et d'action pour une particle
chargée dans un champ électromagnetique oscillant.
Compt. Rend. 268B, 607.

Rébut, P. H. and A. Samain (1969b).

Formalism for study of a plasma in the neighborhood of
steady-state equilibrium. Compt. Rend. 268B, 783.

Ribe, F. L. (1969).

Effect of an oscillating quadrupole field on the kink
modes of a screw pinch. Los Alamos Scientific
Laboratory Report LA-4081 MS.

Ribe, F. L. (1969).

Free-boundary solutions for high-beta stellarators
of large aspect ratio. Los Alamos Scientific Laboratory Report LA-4098.

Ribe, F. L. and M. N. Rosenbluth (1970).

Feedback stabilization of a high-β, sharp-boundaried
plasma column with helical fields. These Proceedings.

Rosenbluth, M. N. (1965).

Microinstabilities. Plasma Physics (International
Atomic Energy Agency, Vienna) p. 485.

Rosenbluth M. N., J. L.Johnson, J. M. Greene, and K. E. Weimer
(1969). Stability limitations for stellarators with sharp
surfaces. Phys. Fluids 12, 726.

Rosenbluth M. N. and R. F. Post (1965).
High frequency electrostatic plasma instability inherent
to "loss-cone" particle distributions. Phys. Fluids
8, 547.

Rosenbluth, M. N., N. Rostoker, and N. A. Krall (1962).
Finite Larmor radius stabilization of "weakly" unstable
confined plasmas. Nucl. Fusion Suppl. Pt. 1, 143.

Rowberg, R. E. and A. Y. Wong (1970).
Collisional drift waves in the linear regime. Phys.
Fluids 13, 661.

Rudakov, L. I. and R. Z. Sagdeev (1961).
Instability of a non-uniform rarefied plasma in a strong
magnetic field. Dokl. Akad. Nauk 138, 581. Sov.
Phys.--Dokl. 6, 415.

Rynn, N. (1964).
Improved quiescent plasma source. Rev. Sci. Instr.
35, 40.

Sagdeev, R. Z. and A. A. Galeev (1969).
Nonlinear Plasma Theory (W. A. Benjamin, Inc.,
New York).

Samain, A. (1969).
Determination of the stationary states of a plasma in
an oscillating electromagnetic field. Compt. Rend.
268B, 1009.

Sato, M. (1970).
Evolution from linear to nonlinear saturation of in-
stabilities and change in zeroth-order conditions.
Phys. Rev. Letters 24, 998.

344

Schuurmann, W., C. Bobeldijk, and R. F. de Vries (1969).
  Stability of the screw pinch. Plasma Physics $\underline{11}$, 495.

Shafranov, V. D. (1956).
  Stability of a cylindrical gaseous conductor in a magnetic
  field. Atom. Energ. $\underline{1}$, 38. J. Nucl. Energy II, $\underline{5}$,
  86.

Shafranov, V. D. (1963).
  Equilibrium of a toroidal plasma in a magnetic field.
  J. Nucl. Energy Pt. C $\underline{5}$, 251.

Shafranov, V. D. (1969).
  Hydromagnetic stability of a current carrying plasma
  column in a strong longitudinal magnetic field.
  Kurchatov Institute of Atomic Energy Report IAE-1853.
  Trans. Culham Laboratory Report CTO/661.

Simonen, T. C., T. K. Chu, and H. W. Hendel (1969).
  Feedback control of collisional drift waves by modulated parallel-electron current sink-experiment and
  interpretation. Phys. Rev. Letters $\underline{23}$, 568.

Stefaňov, R. (1970).
  Thermal diffusion in a highly-ionized plasma column.
  Princeton University Plasma Physics Laboratory
  Report MATT-758.

Stix, T. H. (1960).
  Absorption of plasma waves. Phys. Fluids $\underline{3}$, 19.

Stix, T. H. (1969).
  Finite-amplitude collisional drift waves. Phys. Fluids
  $\underline{12}$, 627.

Strijland, W. (1969).

> Electron containment in a cusped geometry. Instituut voor Plasma-Fysica, Rijnhuizen, Jutphaas, Report 69-53.

Takayama, K., M. Otsuka, Y. Tanaka, K. Ishii, and Y. Kubota (1967). A fully ionized high density plasma source of the TP-D. Eighth International Conference on Phenomena in Ionized Gases (Springer-Verlag, Vienna) p. 551.

Taylor, J. B. and C. N. Lashmore-Davies (1970).

> Plasma stabilization by feedback. Phys. Rev. Letters 24, 1340.

Thomas, R. H. (1967).

> Stability of a distributed parameter system controlled by spatially and temporally sampled feedback. Master's Thesis, M.I.T., Cambridge.

Timofeev, A. V. and V. I. Pistunovich (1970).

> Cyclotron instabilities in an anisotropic plasma. Reviews of Plasma Physics (Consultants Bureau Enterprises, Inc., New York) Ed. M. A. Leontovich, Vol. 5, p. 401.

Trivelpiece, A. W., R. E. Pechacek, and Cj. A. Kapetanakos (1968). Trapping of a 0.5-MeV electron ring in a 15-G pulsed magnetic mirror. Phys. Rev. Letters 21, 1436.

Troyon, F. (1967a).

> Instabilities in a traveling periodic bumpy theta pinch. Phys. Rev. Letters 19, 1963.

Troyon, F. (1967b).

> Stability of a dense plasma confined by a rotating magnetic field. Phys. Fluids 10, 2660.

346

Tsai, S.T., F. W. Perkins, and T. H. Stix (1970).
    Thermal conductivity and low frequency waves in
    collisional plasmas. Plasma Physics Laboratory
    Report MATT-700.

Van der Pol, B. (1934).
    Nonlinear theory of electric oscillations. Proc. I.R.E.
    $\underline{22}$, 1051.

Wang, P. K. C. (1969).
    Optimal control of a class of linear symmetric hyperbolic
    systems with applications to plasma confinement.
    J. Math Anal. Appl. $\underline{28}$, 594.

Wang, P. K. C. (1970).
    Feedback stabilization of highly conducting plasmas.
    Phys. Rev. Letters $\underline{24}$, 362.

Wang, P. K. C. and W. A. Janos (1970).
    A control-theoretic approach to the plasma confinement
    problem. J. Optim. Theor. Appl. $\underline{5}$, 313.

Watson, C. J. H. and L. G. Kuo-Petravic (1968).
    Charged-particle containment in rf-supplemented
    magnetic mirror machines. Phys. Rev. Letters $\underline{20}$, 1231.

Weibel, E. S. (1960).
    Dynamic stabilization of a plasma column. Phys.
    Fluids $\underline{3}$, 946.

Weitzner, H. (1969).
    High beta equilibrium with large helical wavelength.
    Bull. Am. Phys. Soc. $\underline{14}$, 1049.

Wobig, H. and H. Tasso (1970).

    On dynamical stabilization of MHD-instabilities.

    These Proceedings.

Wolf, G. H. (1969a).

    The dynamic stabilization of the Rayleigh-Taylor insta-

    bility and the corresponding dynamic equilibrium.

    Z. Physik 227, 291.

Wolf, G. H. (1969b).

    Use of the M- and S configuration in theta pinches.

    Z. Naturf. 24a, 998.

Wolf, G. H. (1970a).

    Dynamic stabilization of the interchange instability of a

    liquid-gas interface. Phys. Rev. Letters 24, 444.

Wolf, G. H. (1970b).

    Dynamic stabilization of hydrodynamic interchange

    instabilities--A model for plasma physics. These

    Proceedings.

Wolf, G. H. and G. Berge (1969).

    Estimate of a toroidal magnetohydrodynamically stable

    high-β dynamic equilibrium. Phys. Rev. Letters 22,

    1096.

Wong, A. Y. (1965).

    Local heating of a Cs plasma by microwave irradiation

    at the upper hybrid frequency. Appl. Phys. Letters

    6, 147.

Wong, A. Y., D. R. Baker, and N. Booth (1970).

    Efficient modulation coupling between electron and ion

    resonances in magnetoactive plasmas. Phys. Rev.

    Letters 24, 804.

348

Wong, A.Y., M. V. Goldman, F. Hai, and R. Rowberg (1968).
  Parametric excitation from thermal fluctuation at plasma-
  drift wave frequencies. Phys. Rev. Letters <u>21</u>, 518.

Wong, A. Y. and F. Hai (1969).
  Direct measurements of linear growth rates and nonlinear
  saturation coefficients of instabilities. Phys. Rev.
  Letters <u>23</u>, 163.

Wong, A. Y. and A. F. Kuckes (1964).
  Observation of microwave radiation from plasma
  oscillations at the upper hybrid frequency. Phys. Rev.
  Letters <u>13</u>, 306.

Yoshikawa, S. (1964).
  Application of the virial theorem to equilibria of toroidal
  plasmas. Phys. Fluids <u>7</u>, 278.

AUTHOR INDEX

352

Kachalov, O. N.	189
Kadomtsev, B. B.	147, 219
Kapetanakos, Cj. A.	315
Karr, H. J.	274
Kaufmann, M.	81
Kaw, P. K.	226
Kawabe, T.	226, 265
Kawamura, T.	315
Keen, B. E.	1, 25, 103, 105, 110, 112, 114, 119, 120, 121, 123, 128, 129 138, 156, 170, 200, 250, 252
Kent, G. I.	20, 140, 142
Kirillov, V. D.	74
Klochkov, N. I.	189
Kojima, S.	264
Konyaev, V. P.	189
Kornilov, E. A.	264, 265
Kotowski, G.	294, 297
Krall, N. A.	115
Kruskal, M. K.	82
Kubo, R.	266
Kubota, Y.	255
Kuckes, A. F.	90, 95
Kuehl, H.	34
Kulsrud, R. M.	82
Kuo-Petravic, L. G.	310
Kuznetsov, V. V.	74, 189
Landau, L. D.	23, 120, 160, 269, 294
Lashinsky, H.	103, 140, 156, 199, 200, 203, 269
Lashmore-Davies, C.N.	23, 27, 28, 31, 121, 187
Lauer, E. J.	176
Layman, R. W.	266
Lazar, N. H.	315
Lee, D. A.	150
Lepechinsky, D.	214, 216, 219, 220
Levine, A. M.	210, 260, 263
Lewis,P. B.	266
Lifshitz, E. M.	23, 120, 160, 269, 294
Lighthill, M. S.	159
Likhtenstein, V. Kh.	119, 128, 138, 170, 188, 189
Lindgren, N. E.	114, 119, 129
Lindman, E. L.	17, 138
Lisitano, G.	110
Little, E. M.	81, 286
Little, P. F.	114
Lord Rayleigh	294, 300
Lotz, W.	245

SUBJECT INDEX